WITHDRAWN
UTSA LIBRARIES

RENEWALS 458-4574

Underground Storage Systems

LEAK DETECTION and MONITORING

Foreword by Rudolph C. White

Todd G. Schwendeman
H. Kendall Wilcox

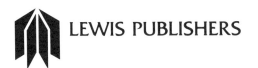

Library of Congress Cataloging in Publication Data

Schwendeman, T. G.
  Underground storage systems.

  Bibliography: p.
  Includes index.
  1. Petroleum products—Underground storage.
  2. Leak detectors. I. Wilcox, H. Kendall. II. Title.
  TP692.5.S36   1987   665.5'42   87-3661
  ISBN 0-87371-045-2

**Library
University of Texas
at San Antonio**

Second Printing 1988

COPYRIGHT © 1987 by LEWIS PUBLISHERS, INC.
ALL RIGHTS RESERVED

Neither this book nor any part may be reproduced or transmitted in any form or by any means, electronic or mechanical, including photocopying, microfilming, and recording, or by any information storage and retrieval system, without permission in writing from the publisher.

LEWIS PUBLISHERS, INC.
121 South Main Street, Chelsea, Michigan 48118

PRINTED IN THE UNITED STATES OF AMERICA

To my mother and father

This book is a tribute to their unfailing encouragement and devotion

                                        Todd Schwendeman

To my parents,

who supported me in continuing my education

                                                Ken Wilcox

Foreword

When President Reagan signed the Hazardous and Solid Waste Amendments to the Resource Conservation and Recovery Act (RCRA) on November 8, 1984, national attention was focused on underground storage tanks. The new Subtitle I of RCRA directs the Environmental Protection Agency (EPA) to issue regulations concerning all aspects of underground storage tank management.

Just a few short years ago, few individuals were concerned with the universe of underground storage tanks. The only people involved with the management of these tanks were the manufacturers and users. The petroleum industry has naturally been involved with underground storage tank management through much of its history. The American Petroleum Institute (API) has had an active underground storage tank program since the 1960s, a program designed to prevent leaks by the selection and correct installation of the proper tanks and associated piping. The program is also designed to educate tank owners and operators concerning procedures for detecting leaks from underground storage tanks. These procedures include investigation to determine the presence of a leak and methods to secure and clean up leaked petroleum products.

Not too many years ago, the only leak monitoring systems available for underground storage tanks were inventory monitoring of the products and line leak detectors for submersible pumps. When either of these systems indicated a leak, or unconfined product was detected by other means, the storage tank was tested for tightness.

During the past few years, there has been an explosive growth of methods to detect and monitor for leaks from underground storage tanks. During the past five to six years, more devices and methods have been developed for monitoring underground storage tanks than were developed in the previous 50 years. This book is designed to

---

Note: The views expressed in this foreword are entirely my own, and do not necessarily represent the views of the American Petroleum Institute. Mention of particular techniques or devices does not constitute endorsement by the American Petroleum Institute, nor does it constitute endorsement by the author.

acquaint underground storage tank owners and operators with the many techniques and devices available today for monitoring underground storage tanks.

We in the petroleum industry believe that inventory monitoring of the product is critical to the early detection of leaks. Historically, manual inventory has proven to be of considerable value to the petroleum industry by providing the first indication of a product loss. When inventory records indicate a product loss, a series of steps should be taken as a follow-up to determine if the tank is actually leaking. These steps are described in API Publication 1621, "Recommended Practices for Bulk Liquid Stock Control at Retail Outlets" (1977).

A number of automatic tank gauging devices have been developed recently and are available as "off the shelf" equipment. This equipment measures the height of liquid in a storage tank and uses a microprocessor to convert the height into a volume measurement. The devices can also be operated in a leak detection mode, and most claim a leak detection capability of 0.2 gal/hr. These devices may prove to be of great value in the future in eliminating the "human element" from inventory control procedures, thus providing for automation of product inventory control and leak detection in underground storage tanks.

Leak detection for underground storage tanks is normally divided into systems located outside the tank and systems located within the tank. Automatic tank gauges, as previously noted, are an example of an in-tank leak detection system. An example of external systems is the use of observation wells outside the tank and within the backfill area of the tank field to detect leaks. This book describes the construction, location, and sampling of observation wells. The wells must be properly constructed and located close to the tank if they are to detect underground leaks rapidly and effectively. The wells may be manually sampled, or a continuous automatic leak detector may be placed in the wells to detect petroleum hydrocarbons. The authors describe the types of automatic leak detectors available and the advantages and disadvantages of each. Observation wells are normally effective where groundwater is relatively close to the surface (25-40 ft). With new tank systems, U-tube observation wells may be useful to detect leaks at locations where groundwater levels are far from the surface.

Vapor monitoring systems may also prove to be useful under conditions where groundwater is located far below the surface. These systems can detect released products, such as motor fuels, that have appreciable vapor pressures. Normally, one or several sample-gathering devices are placed in the vadose zone above the groundwater table. The detection devices may be located directly in the sample collectors or a means can be provided to collect the sample and transport it to the detection device. A number of detection devices are available that will detect very low concentrations of hydrocarbon vapors. These devices have been used to detect leaks from underground storage systems only within the last several years, and their reliability needs to be demonstrated to regulatory agencies and the tank user industry.

EPA has recently published regulations for hazardous waste tanks. They call for the use of double-walled tanks as a means for secondary containment. The regulations further call for interstitial monitoring of the double-walled tanks. This book describes the methods available for interstitial monitoring. Some of the systems are specifically designed for double-walled tanks, and many of the systems described for leak detection in observation wells can also be used for detecting leaks in the interstitial space of double-walled tanks. The authors also point out that manual interstitial monitoring can be an effective leak detection technique.

However, because of the varied soil, geological, and groundwater conditions in the United States, it is difficult to justify universal use of doubled-walled tanks. In many areas, the proper installation of corrosion-resistant single-walled tanks (as called for under the RCRA Interim Prohibition Regulations) with the appropriate monitoring systems will provide more than adequate environmental protection for leaks from underground tanks. Double-walled tanks and other secondary containment systems should be required only in environmentally sensitive areas, such as in the cones of depression of municipal wells, where it can be demonstrated that sole-source aquifers require such protection, or where the tanks are to be installed close to drinking water wells. Regardless of the means of secondary containment used, a means for monitoring for a leak from the system should be provided, whether the monitoring is manual or automatic.

The authors have devoted a chapter to the testing and monitoring of the piping systems associated with underground tanks. The piping systems are an important part of underground tank systems and, depending upon the source quoted, can be responsible for up to 60% of the underground leaks reported. It makes eminent sense that piping would be responsible for a large number of leaks when one considers the extensive piping system and the number of joints, elbows, and unions required for the system. As the authors point out, means exist for appropriately testing and monitoring piping systems. Hopefully, more accurate and reliable systems for monitoring piping will emerge in future years as more attention is devoted to this aspect of underground product leak prevention.

Monitoring systems for detecting leaks are extensively covered in this book, but the authors have not neglected underground tank testing methodologies. It is important to distinguish monitoring systems from tank testing procedures, since both have a definite role to play in the proper management of underground tanks. Monitoring techniques provide a means to continuously or semicontinuously determine whether a system is leaking, whereas tank testing provides a means for determining at the time of the test whether a leak is occurring. The authors point out that the accuracy of existing tank testing procedures is limited by many variables.

Tank testing has been used for many years by the petroleum industry to confirm whether a system is leaking. Because of the attention being directed to tank testing procedures by regulatory agencies and tank owners and operators, the testing procedures will probably become more sophisticated by the application of more refined equipment, including, probably, more automation.

One cannot read this book without concluding that there are a bewildering number of combinations for the proper management of underground tanks. Decisions must be made concerning the proper tank to use, the type of piping, the correct monitoring system, and the proper system for a particular location. These decisions require considerable knowledge of the systems available. The authors have designed this book as a convenient reference source for installers, users, and owners of underground tanks, providing instructions on what is available and how it can be best used to design the tank system most appropriate for a particular location. The authors have succeeded in combining the available information into one concise document that will greatly assist the tank owner and operator.

Electronic devices and automatic equipment can never completely replace the diligence and responsibility required of the owners and operators of storage systems. Owners and operators have a responsibility to manage their underground tank systems using due diligence, regardless of whether simple manual techniques or sophisticated devices are used. Conscientious attention to the operation of an underground storage system is a key responsibility for storing and dispensing hazardous substances.

>Rudy White
>Manager, Market Operations
> and Engineering
>American Petroleum Institute

Preface

The objective of *Underground Storage Systems: Leak Detection and Monitoring* is to provide owners and operators of underground storage systems with in-depth technical information on the technologies available for monitoring for leaks and losses. The text reviews the various types of monitoring equipment and techniques available and how they can be integrated to achieve a comprehensive monitoring program.

This book is a valuable reference for storage system owners, fleet operators, engineers, geologists, environmental scientists, petroleum distributors, maintenance staff, facility planners and designers, and regulatory officials. It is designed to address the principal methodologies of underground storage system leak detection and monitoring: inventory monitoring, external and internal tank technologies, and piping technologies.

Todd G. Schwendeman
H. Kendall Wilcox

Acknowledgments

As with any effort of this magnitude, the authors' debts are considerable and wide ranging.

We especially express our gratitude for valued historical and technical information and inspiration to Drs. William Deaver and J. T. Dibble at the Texaco Research Laboratory in Port Arthur, Texas.

We would also like to thank the American Petroleum Institute's Underground Storage/Groundwater Task Force, including specifically: Sully Curran, Exxon; Jim Rocco, Sohio; Ken Smith, Charlie Van Inwagen, Ted Kirkpatrick (retired), Shell; James Krantzthor, Chevron; Denny Strock, Amoco; Don Hitchcock, Matty Mattson (retired), Dick Kregel, Texaco; Gary Weiland, Marathon; Lem McMannes (retired), Marathon; Jim White, ARCO; Bill Beck, Mobil; Carl Hasselback, White Arrow Service Stations; and the many other participating members.

Many helpful discussions were held with colleagues regarding the material in Chapter 4. Dr. Jairus Flora and Dr. Clarence Haile, Midwest Research Institute; Dr. Joseph Maresca, Vista Research Institute; Mr. William Purpora, Protanic, Inc.; Mr. Mike Kalinoski, U.S. EPA; and Dr. Glenn Thompson, Tracer Research Corporation, provided many valuable insights and important technical information.

The authors extend a sincere thanks to Dr. Rudy White of the American Petroleum Institute for his overall guidance in organizing the text and for contributing the foreword.

Many editors provided comments on this book. Dr. Bruce Bauman, of the American Petroleum Institute, and Paul Yaniga and Michael Brenoel, of Groundwater Technology, Inc., require recognition for their efforts to improve the structure and technical quality of the manuscript. The willing dedication of their time and efforts is greatly appreciated.

Joanne Detweiler and Leslie Ross of Groundwater Technology, Inc., deserve a special tribute for their secretarial skills and fine graphic arts ability, respectively. They performed above and beyond the call of duty. Many special thanks are also extended to Doris Nagel for the excellent technical editing on Chapter 4.

Last but not least, the authors would like to thank Terry Schwendeman and Kathy Wilcox, whose support, patience, and understanding were instrumental in helping them achieve the completion of the manuscript.

**Todd G. Schwendeman** is Director of Storage Systems Management Services for the Annapolis Junction, Maryland office of Groundwater Technology, Inc. He develops and implements comprehensive storage system management programs that include risk assessment, release detection and monitoring, inventory monitoring, storage system closure and replacement, release remediation, and staff training.

Formerly, Mr. Schwendeman was the Underground Storage Systems Coordinator for the American Petroleum Institute, where he was in charge of the petroleum industry's technical response to legislation and regulation concerning underground storage systems. In 1982–1983 he was employed by Texaco at the Port Arthur Research and Technical Laboratory, investigating and assessing groundwater, surface water, and soils contamination incidents. He has conducted water quality studies for the U.S. Forest Service, and has performed technical analyses for a soils engineering firm.

Mr. Schwendeman holds an MS in civil engineering from Montana State University (1981) and a BS in biology from Colgate University (1978). He is author or coauthor of numerous papers and research reports.

**H. Kendall Wilcox** has been employed by Midwest Research Institute for the past eight years. He has served as project leader on several underground storage tank projects over the past four years. These have included projects for EPA conducting field evaluations of several test methods and conducting a test survey of underground storage tanks across the country. During these projects, Dr. Wilcox conducted numerous tests, evaluated the data, and experienced many of the problems which are typical of underground tank testing.

The combination of field experience, theoretical background, and controlled evaluations has given Dr. Wilcox broad experience covering most of the test methods available today and their application to a wide range of tank designs and sizes.

Dr. Wilcox has been an invited speaker at several technical conferences, including those sponsored by the ASME, the University of Wisconsin, the U.S. Air Force, and the U.S. EPA. He is the author or co-author of numerous technical reports and is a member of the Sigma Xi. Prior to his employment at MRI, he was involved in environmental testing, quality assurance, and teaching. He received his BA in chemistry from Sterling College in 1964 and his PhD in chemistry from the University of Southern California in 1972.

Contents

1. **OVERVIEW** 1
   - Purpose 3
   - Types of Underground Tanks 4
     - Cathodically Protected Steel 4
     - Steel-Clad with Corrosion-Resistant Materials 8
     - Noncorrosive Materials 9
     - Double-Walled 10
   - Types of Pipe and Pumping Systems 11
     - Cathodically Protected Steel 11
     - Noncorrosive Materials 13
     - Double-Walled 13
     - Pumping Systems 13
   - Types of Regulated Substances 14
   - Types of Release Detection and Monitoring Systems 15
   - Summary 16
2. **INVENTORY MONITORING** 19
   - Introduction 21
   - Regulatory Requirements 23
   - Factors Affecting Inventory Monitoring 25
     - Temperature 25
     - Meter Accuracy 27
     - Evaporation 29
     - Gauging Accuracy 30
     - Tank Geometry 31
   - Inventory Monitoring Techniques 32
     - Manual Reconciliation Techniques 32
     - Statistical Reconciliation Techniques 42
     - Automatic Gauging Systems 47
   - Summary 48
3. **EXTERNAL TANK RELEASE DETECTION AND MONITORING** 53
   - Introduction 55
   - Contaminant Transport 57
     - Liquid 57
     - Dissolved-Phase 59
     - Vapor 61

Release Detection Techniques		64
Groundwater Detection Techniques		64
Detection Wells		65
Soil Sampling		73
Dyes and Tracers		73
Surface Geophysics		81
Vadose Zone Vapor Detection Techniques		83
Grab Sampling of Soil Cores		83
Surface Flux Chambers		84
Downhole Flux Chamber		84
Accumulator Systems		86
Ground Probe Testing		88
Physical Inspection		90
Visual Inspection		90
Integrity Testing (Tank Shell)		90
Release Monitoring Systems		93
Observation Wells		94
Design		94
Construction		95
Installation		96
Sampling		97
Vapor Wells		103
Design		104
Construction/Installation		104
Sampling		105
U-Tubes		107
Design		108
Construction/Installation		109
Sampling		109
Secondary Containment Monitoring		109
Summary		111
4	IN-TANK LEAK DETECTION METHODOLOGIES	117
Basic Principles of Tank Testing		120
Volumetric Testing		122
Temperature Compensation Techniques		126
Product Level Measurements		129
Calibration of Level Sensor		134
Determination of the Coefficient of Expansion		135
Variables in Volumetric Testing		136

	Temperature Effects 137
	Vapor Pockets 140
	Water Table Effects 143
	Tank Distortions 144
	Vibration 146
	Evaporation and Condensation 147
	Head Pressure Effects 147
Nonvolumetric Testing 149
Helium Leak Detection 150
Tracer Leak Detection 151
Other Nonvolumetric Methods 153
Continuous In-Tank Monitors as Leak Detectors 153
Interpretation of Results 154
How to Use This Information to Reduce the Risk
of a Bad Test 159
5	PIPING RELEASE DETECTION AND
MONITORING 163
Introduction 165
Pressure Monitoring 166
Positive Pressure 166
Negative Pressure 171
Tightness Testing 171
Direct Testing 172
Indirect Testing 172
External Sensing Systems 173
Vapor Monitoring 174
Cable Systems 174
Containment Technologies 174
Low-Permeability Soils 176
Impervious Barriers 176
Double-Walled Piping 177
Concrete Encasement and Soil Cements 181
Summary 181
APPENDIX A   Hazardous Substance List,
Comprehensive Environmental Response,
Compensation, and Liability Act of 1980, Section
101 (14) 185
APPENDIX B   Storage System Gauging Procedures 203
APPENDIX C   Equipment and Procedure for
Testing the Accuracy of Gasoline Dispensing
Meters 205

APPENDIX D State Underground Storage
  Regulatory Programs: Release Detection and
  Monitoring Requirements in California, Delaware,
  and Florida 207
Index 211

List of Figures

Chapter 1
1	Impressed current cathodic protection	7
2	Sacrificial anode cathodic protection	8
3	Schematic diagram of a remote pumping system	14
4	Schematic diagram of a suction pumping system	15

Chapter 2
1	Temperature effects on inventory monitoring	26
2	Histogram for slow flow meter testing	27
3	Histogram for fast flow meter testing	28
4	Relationship between meter error in gal per gal dispensed and meter error in in.^{3} in a 5-gal test	29
5	Error due to $1/8$ in. gauging stick measurement in a 10,000- and 20,000-gal horizontal tank (8 ft 0 in. diameter)	30
6	Sample daily inventory data entry form	33
7	Sample daily inventory data entry form	34
8	Sample daily inventory data entry form	35
9	Static reconciliation data entry form	36
10	Sample statistical inventory review form – daily inventory	45
11	Sample statistical inventory review form – thermal shrinkage	46

Chapter 3
1	Transport of hydrocarbon in the subsurface	58
2	Capillary zone restricting hydrocarbon movement on the water table	62
3	Contamination of the vadose zone and capillary zone by a fluctuating water table	63
4	Typical detection well	70
5	Typical detection well cluster	72
6	Surface flux chamber and peripheral equipment	86
7	Downhole isolation flux chamber	87
8	Curie-point accumulator device	88
9	Driven probe soil gas sampling system	89

10	Dual-tracer leak detection method	90
11	Typical observation well installations	96
12	Schematic diagram of a bailer	98
13	Schematic diagram of a differential float device	100
14	Schematic diagram of an electrical resistivity-sensing device	101
15	Schematic diagram of a hydrocarbon-soluble electrically activated device	102
16	Vapor well monitoring system	105
17	Schematic diagram of a metal oxide semiconductor	106
18	Schematic diagram of a U-tube installation	108
19	Typical Type 2 secondary containment sensor installation	110

Chapter 4

1	Schematic of a simplified tank test apparatus	124
2	Typical test data illustrating the volume effects of temperature and level as related to the leak rate	125
3	Trend behavior in a tight tank	126
4	Trend behavior in a leaking tank	127
5	Schematic of a device used to monitor changes in volume by maintaining product level at a constant height	129
6	Level changes detected by changes in head pressure	130
7	Level changes detected by change in pressure necessary to produce bubbles	131
8	Use of a buoyancy probe to measure level changes	131
9	Laser interferometer used to measure level changes	132
10	Level changes detected by changes in intensity of light reaching photocell	133
11	Product level behavior during calibration using two volumes	135
12	Temperature changes which will produce a volume change of 0.05 gal as a function of tank size	137

13	Schematic of the location of several vapor pockets in a tilted tank	140
14	The effect of barometric pressure changes on the volume of a vapor pocket	142
15	The effect of temperature changes on the volume of a vapor pocket	143
16	Behavior exhibited by the tank distortion which has not reached equilibrium	145
17	Head pressure effects on leak rate for two holes of undefined size	149
18	Schematic diagram showing dual-tracer leak detection concept	152
19	Noise histograms for a nonleaking and leaking tank, showing probability of detecting false alarms and probability of detecting a leak of 0.10 gph when the criterion is set at 0.05 gph	155
20	Probability of detecting a leak for $C = 0.05$ and values for s of 0.03 and 0.05	156

Chapter 5

1	Mechanical line leak detector	168
2	Electronic positive pressure monitoring system	170
3	Vapor monitoring of a piping system	175
4	Lined storage tank excavation and piping trench	178
5	Double-walled tanks with a lined piping trench	179
6	Double-walled pipe and fittings	180

List of Tables

Chapter 1

1	Thickness of steel tank shells recommended by Underwriters Laboratories	6
2	Typical anode characteristics	8
3	Number of feet of well-coated steel pipe which can be protected with one galvanic anode	12

Chapter 2

1	State inventory monitoring regulatory requirements	24
2	Typical underground steel tank configuration	31
3	Action numbers for static reconciliation	37
4	Action numbers for metered reconciliation	39
5	Example of temperature-compensated inventory monitoring	39
6	Summary of procedures selected for evaluation	41
7	Comparison of the performance of evaluated inventory monitoring procedures	42
8	Comparison of evaluated inventory monitoring procedures by performance group	43
9	Performance characteristics of automatic gauging systems	49

Chapter 3

1	Well drilling methods	66
2	Evaluation of well drilling methods	67
3	Types of groundwater detection wells	71
4	Factors affecting number of wells per location (clusters)	72
5	Advantages and disadvantages of groundwater detection methods	74
6	Suitability of well sampling methods	79
7	Typical ranges of geophysical variables	82
8	Criteria for selecting soil sampling equipment	85
9	Comparison of vadose zone vapor sampling techniques	91

10 Desired properties of an external release
    monitoring system 93

Chapter 4
 1 Typical coefficients of expansion for selected
    materials 136

Appendix D
 1 Schedule for retrofitting of existing underground
    storage tanks, Florida regulations 210

CHAPTER 1

Overview

Todd G. Schwendeman

CHAPTER CONTENTS

Purpose 3
Types of Underground Tanks 4
   Cathodically Protected Steel 4
   Steel-Clad with Corrosion-Resistant Materials 8
   Noncorrosive Materials 9
   Double-Walled Tanks 10
Types of Pipes and Fittings 11
   Cathodically Protected Steel 11
   Noncorrosive Materials 13
   Double-Walled Piping 13
Types of Pumping Systems 13
Types of Regulated Substances 14
Types of Release Detection and Monitoring Systems 15
Summary 16

Overview

Todd G. Schwendeman

PURPOSE

Since the early 1970s, the United States has experienced sweeping environmental reforms. Increased scientific information on the toxicity and long-term health effects of many compounds has resulted in fundamental changes in the handling and use of hazardous materials. And as the nation continues to expand, increased demands are being placed on our valuable natural resources. In particular, the use of groundwater as a drinking water source has increased threefold since the 1950s, with no indication that this trend will decline in the future.

These two factors, along with a better understanding of the fate and effects of hazardous materials transport in the subsurface environment, are causing major modifications in the way substances are stored. Underground storage system owners and managers are confronted with skyrocketing economic liability, rapidly increasing insurance costs (if coverage is even obtainable), and the implementation of complex government regulatory programs. The owner or manager must justify the convenience of storing regulated hazardous substances onsite, and must examine closely the benefits of retailing regulated substances. The decision to continue to store regulated substances onsite or to remove or abandon storage systems will have a significant impact on most business operations in the future. (The storage of hazardous substances in the subsurface has in most situations been mandated as a result of fire and safety concerns.)

Underground Storage Systems: Leak Detection and Monitoring has been prepared to assist the owner or manager of an underground storage system in evaluating the options currently available to identify or monitor for unconfined stored substances. The purpose of the book is to provide an owner or manager with an overview of the equipment and techniques currently available to manage underground storage systems better. Although definitive data are not available on the number, type, and size of underground storage systems in the United States, most systems do contain petroleum products, and the discussion throughout this text will focus on petroleum products.

However, an effort will be made to identify those situations or techniques that do not universally apply to materials stored beneath the ground.

The term underground storage system, as used in this text, includes the tank itself, piping, fittings, the pumping system, and all associated appurtenances. A release can occur from any of the liquid-handling components of an underground storage system; therefore, the entire system must be managed critically to minimize the potential of a release.

The information contained in this text has been compiled from textbooks, technical journals, guidance manuals, regulatory codes, and many personal communications with individuals knowledgeable in the field. An effort has been made throughout the text to describe the various release detection and monitoring technologies without describing specific types of equipment.

TYPES OF UNDERGROUND TANKS

In this section, different types of designs and materials for storage tanks will be briefly described, and the advantages of each system will be detailed. Until the early 1970s, steel was the principal construction material for underground storage systems. A limited number of systems were constructed of concrete; these are located principally at Department of Defense facilities and are classified as Category I tanks, usually ranging in size from 20,000 to 50,000 barrels (a barrel is 42 gal), or Category II tanks, containing 50,000 to 80,000 barrels of fuel. The following kinds of tanks will be described: cathodically protected steel, noncorrosive or corrosion-resistant materials, steel-clad with corrosion-resistant materials, and double-walled.

Cathodically Protected Steel

As a first step in preventing releases from underground storage systems, the Resource Conservation and Recovery Act Amendments of 1984 restrict the types of tanks that can be installed to store hazardous substances. An underground storage tank conforming to the Interim Prohibition, which applies until final rules are promulgated, is designed to "prevent releases due to corrosion or structural failure for the operational life of the tank; is cathodically protected against corrosion, constructed of noncorrosive material, or designed in a manner to prevent the release or threatened release of any stored substance; and the material used in the construction or lining of the tank is compatible with the substance to be stored." A bare steel tank (no cathodic protection) can be installed only at those sites in which the soil has a resistivity greater than 12,000 ohm/cm as determined by American Society for Testing and Materials (ASTM) Standard G57-78. Although the statutory language is broad, a clear statement has been made that storage systems which fail to provide a high level of environmental protection are no longer acceptable.

The fabrication and design requirements for steel underground storage tanks are specified in Underwriters Laboratories (UL) Standard 58, entitled "Standards for Steel Underground Tanks for Flammable and Combustible Liquids."[1] The requirements of the American Society of Mechanical Engineers (ASME) are detailed in the ASME Pressure Vessel Code, Section VIII, Division I, Boiler and Pressure Vessel Code.[2] This manufacturing requirement stipulates that the length of the tank not exceed six times the diameter. The minimum shell thicknesses required by UL for tanks up to 50,000 gal in capacity are illustrated in Table 1. Galvanized steel is considered a suitable fabrication material for tanks smaller than 1,100 gal and these tanks are permitted to have a shell thickness between 0.07 and 0.14 in. Butt welding or lap welding is permitted according to UL or ASME standards, although welded butt joints are stronger and less susceptible to corrosion than lap joints.

Corrosion of underground steel is a natural phenomenon governed by electrochemical principles. Galvanic corrosion can be defined as an electrochemical degradation of a metal as a result of a reaction with its environment.[3] Corrosion protection can be achieved by using either an impressed current system, which incorporates direct current and an energized anode to deliver current to the tank, or a galvanic anode system, which relies on the natural potential difference between the metallic sacrificial anode and the steel tank to provide a protective flow of current. Protection of the tank is achieved when a structure-to-soil potential of −0.85 V, as measured by a copper–copper sulfate reference half cell, is attained.

The impressed current system is most often applied to storage systems with large areas of exposed steel. It is commonly used to protect existing tanks and should be designed and installed by a qualified corrosion engineer or by someone knowledgeable in the science of corrosion. Figure 1 illustrates an impressed current protection system. A galvanic or sacrificial anode system is typically restricted to new

Table 1. Thickness of Steel Tank Shells Recommended by Underwriters Laboratories[a]

Capacity		Maximum[b] Diameter		Manufacturers Standard or Galvanized Sheet	Nominal Thickness			
					Uncoated		Galvanized[c]	
U.S. Gallons	dm3	Inches	m	Gauge No.	Inches	mm	Inches	mm
Up to 285	Up to 1,078	42	1.07	14[c]	0.075	1.91	0.079	2.01
286 to 569	1,082 to 2,120	48	1.22	12[c]	0.105	2.67	0.108	2.74
561 to 1,100	2,124 to 4,164	64	1.63	10[c]	0.135	3.43	0.138	3.51
1,101 to 4,000	4,168 to 15,142	84	2.13	7	0.179	4.55		
4,001 to 12,000	15,145 to 45,425	126	3.20	1/4 inch	0.250	6.35		
12,001 to 20,000	45,429 to 75,708	144	3.66	5/16 inch	0.312	7.92		
20,001 to 50,000	75,712 to 189,270	144	3.66	3/8 inch	0.375	9.53		

[a]Source: Underwriters Laboratories 58.[1]
[b]Length of tank shall not be greater than six times the diameter.
[c]The use of galvanized steel and shell thicknesses less than 7 gauge is not recommended by the New York Department of Environmental Conservation.

OVERVIEW 7

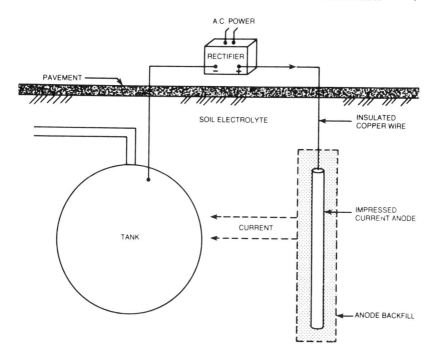

Figure 1. Impressed current cathodic protection. (Source: American Petroleum Institute Publication 1632.)[4]

tanks in the form of a preengineered system that includes a coating over the outside of the tank and electrical isolation of the tank from the piping. Figure 2 illustrates a sacrificial anode protection system. The coating reduces the surface area of the tank that is directly exposed to a corrosive environment, while electrical isolation prevents stray electrical currents from reducing the effectiveness of the sacrificial anode protection.

Three sacrificial anode materials are commonly applied to protect underground storage systems from corrosion: high-potential magnesium alloy, standard magnesium alloy, and zinc.[4] These anodes are usually packaged in low-resistivity (50 ohm/cm) backfill composed of hydrated gypsum, bentonite, and sodium sulfate. Low-resistance backfill reduces anode-to-earth resistance, stabilizes the anode's potential, and improves the efficiency of the anode. Table 2 is an overview of typical anode characteristics.

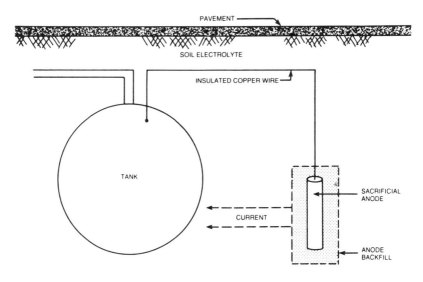

Figure 2. Sacrificial anode cathodic protection. (Source: American Petroleum Institute Publication 1632.)[4]

Steel-Clad with Corrosion-Resistant Materials

Steel tanks clad with a corrosion-resistant material consist of a basic steel tank (typically manufactured to UL 58 specifications) which has an exterior coating, approximately 100 mm thick, of a corrosion-resistant material. Fiberglass and coal-tar epoxy are common exterior coatings. Underwriters Laboratories has a specific listing for steel tanks clad with a corrosion-resistant material, but an official standard has not been issued. This type of tank, commonly referred to as a composite tank, combines the structural properties of steel with a corrosion-resistant external coating.

To apply the coating for a composite tank, the outer shell first must

Table 2. Typical Anode Characteristics[a]

Anode Data	Magnesium	Zinc
Theoretical current capacity, amp hours per pound	1,000	372
Current efficiency	50%	90%
Actual consumption rate, pounds per amp year	17	24.8
Solution potential to a copper–copper sulfate cell	−1.55	−1.1
(Same) for high potential magnesium	−1.73	

[a]Source: Petroleum Equipment Institute RP 100–86.[8]

be sandblasted free of rust and scale using the Steel Structures Painting Council specifications for commercial blast cleaning, SSPC-SP-6. Once the coating has cured, both visual inspection and electrostatic testing with 10,000 V are used to detect coating imperfections. (After a tank leaves the factory, damage to the coating may occur during transport or installation.) This procedure is commonly referred to as "spark testing." A quality coating must also possess high bonding qualities and have a coefficient of expansion compatible with that of steel. To prevent stray-current corrosion, dielectric bushings should be used to electrically isolate the tank from the piping system.

Noncorrosive Materials

The most commonly used noncorrosive material for the construction of underground storage tanks is fiberglass-reinforced plastic (FRP). FRP tanks should be UL listed and designed in accordance with Underwriters Laboratories Standard 1316, "Standard for Glass-Fiber-Reinforced Plastic Underground Storage Tanks for Petroleum Products"[5] or Underwriters Laboratories of Canada Standard ULC-S615-1977, "Standard for Reinforced Plastic Underground Tanks for Petroleum Storage."[6] The UL specifications are intended to provide the manufacturer with construction guidelines. These specifications address such subjects as requirements for structural strength, criteria for chemical compatibility of tank material with the stored substance and surrounding soil, and test methods for quality control and quality assurance.

To improve the structural strength of FRP tanks, reinforcing ribs are commonly incorporated into the design. Ribs are intended to withstand the internal stresses from the stored substance as well as the external backfill, surface-loading, hydrostatic, and buoyancy forces. Since FRP is more brittle than steel, FRP tanks must be handled and installed more carefully than steel tanks. To assure a sound, high-integrity installation, the installer should adhere carefully to the manufacturers' instructions.

The compatibility of FRP and the stored material should be carefully considered. Many resins are available for use in FRP and each has its own characteristics of performance. To assure that a specific resin will maintain a high level of performance when in contact with a substance, immersion testing should be conducted. Immersion testing should consider bonding strength, flexural strength, impact resistance, Barcol hardness, and film integrity. Compatibility of the stored substance with FRP should be carefully considered, particularly if thought is being given to switching substances from tank to tank or if an FRP tank of unknown age and origin is to be placed into new service.

To address the compatibility issue, three types of UL listed FRP underground storage tanks are currently available: (1) for petroleum products only; (2) for petroleum products mixed with a maximum of 10% percent ethyl alcohol; and (3) for 100% alcohol (including petroleum products with alcohol concentrations in excess of 10%).

Double-Walled Tanks

Double-walled tanks have only recently been introduced into the United States, although they have been in service in Europe for several years. A double-walled design generally consists of a heavier-gauge inner tank surrounded by a lighter-gauge outer shell that permits the monitoring of the interstitial space between the two shells. Interstitial monitoring will be discussed in greater detail in Chapter 5.

Underwriters Laboratories currently lists two specific double-walled designs, Type 1 and Type 2. The Type 1 design consists of a primary tank wrapped by an exterior steel shell that is in direct contact with the primary container. The exterior shell may wrap less than the full 360 degrees of the primary tank's circumference. The Type 2 tank is designed so that the outer shell is physically separated from the inner shell by spacers or standoffs. The inner and outer shells are each a full 360-degree cylinder. The Type 1 tank is extremely rigid, whereas a Type 2 tank requires a much more sophisticated design for the primary shell, since the backfill provides no support to this portion of the tank. This inner tank must be constructed according to UL 142 "Steel Aboveground Tanks for Flammable and Combustible Liquids" specifications. Monitoring of the interstitial space can be conducted for both the Type 1 and Type 2 tanks. To date, UL has not formally published double-walled specifications.

A number of the newer double-walled tank designs consolidate the piping connections in one tank-top location. By introducing all fittings into one manhole, the design offers a number of advantages, including improved control over tightness testing, improved tank and piping cathodic protection monitoring, and built-in overfill protection. Connection of fittings to this type of tank can be more complicated because of limited working space.

TYPES OF PIPES AND FITTINGS

Pipes installed to convey liquids from underground storage tanks to a dispenser, a process unit, or another tank should adhere to environmental requirements similar to those that apply to storage tanks. It is well established that piping systems are a primary source of leakage from underground storage systems. Considerations in piping system design include:[7]

- the corrosive characteristics of the material to be transported, and the ability of the piping system components to withstand that corrosion
- the characteristics of the external atmosphere (soil, water, etc.) to which the piping system components are exposed
- force loadings due to dynamic effects (fluid flow) and weight considerations
- the susceptibility of the piping material to crevice corrosion in threaded joints, in socket-welded joints, and in other stagnant, confined areas
- the possibility of adverse electrolytic effects if the metal is subject to contact with a dissimilar metal, including older metals of the same type
- the chemical compatibility between lubricants or sealants used on threads and the fluid handled
- the suitability of packing seals and O-rings for the fluid service, including possible effects of or on the fluid handled
- the compatibility of materials such as cements, solvents, solders, and brazing with the fluid being transported

Cathodically Protected Steel

Adequate corrosion protection for bare steel pipe can be readily achieved by either sacrificial anode or impressed current cathodic protection. The protection principles that apply to tanks also apply to piping. When anodes are used, the piping should be electrically isolated from the tank and well-coated to reduce contact with the environment. The intent of wrapping or coating pipe is to provide a durable protective coating that will prevent direct contact of the metal with the soil or soil moisture. Pipe coating materials can be painted, brushed, or sprayed on the outer surface of the pipe. Wrapping pipe with a vinyl-coated, mastic-backed wrapping tape is also effective. It is important to note that under the requirements of the EPA Interim Prohibition, as well as many state underground storage regulations,

Table 3. Number of Feet of Well-Coated Steel Pipe Which Can Be Protected with One Galvanic Anode

Anode Weight:	ASTM AZ63,Type II Magnesium Anode			Zinc ASTM B418-73 Type II Anode		
	9#	17#	32#	5#	30#	50#
Corrosive soil, 2,000 ohm cm soil						
2" pipe		1,000+	1,000+	140	450	410
3" pipe		830	1,000+	90	310	280
4" pipe		630	750	70	240	240
Anode life, years		10	16	17	28	50+
Mildly corrosive soil, 5,000 ohm cm soil						
2" pipe	350	530	630	50	180	160
3" pipe	230	360	420	40	120	110
4" pipe	170	280	330	30	90	80
Anode life, years	19	23	37	43	50+	50+
Slightly corrosive soil, 15,000 ohm cm soil[2]						
2" pipe	100	180	200	20	50	50
3" pipe	70	120	140	10	30	40
4" pipe	50	100	110	10	20	30
Anode life, years[3]	50+	50+	50+	50+	50+	50+

[a]Source: Petroleum Equipment Institute, RP 100–86.[8]

the use of galvanized pipe without any additional corrosion protection is considered to be unsatisfactory protection. As a guide to determine the proper amount of sacrificial anode cathodic protection to a piping system, the Petroleum Equipment Institute has developed a useful table that covers a series of anode sizes over a range of different soils.[7] The figures in Table 3 reflect a safety factor of 20–30%, with calculations based on well-coated pipe with less than 5% exposed steel evenly distributed along the pipe and current density of 1.5–2.0 mA per square foot of exposed steel.

An impressed current protection system, most commonly applied to existing systems to prevent any further corrosion, is designed to protect the entire underground storage system, including tanks and piping. Impressed current protection employs direct current from an external source and nonsacrificial anodes. Increased maintenance attention should be devoted to this system to assure that it is functioning properly.

In either type of cathodic protection system, assistance from a corrosion engineer or an individual knowledgeable in the field of corrosion protection is recommended. Technical guidance is available from the National Association of Corrosion Engineers or publications such as API Publication 1632.

Noncorrosive Materials

As with noncorrosive tanks, FRP is the principal noncorrosive material used for piping. FRP piping provides a flexible system that can accommodate minor misalignment and earth movement. The reduced need for swing joints has resulted in increased acceptance of this material for underground storage distribution systems. However, as with FRP tanks, the compatibility of the stored substances with the FRP resin composition must be considered. Considerable care must be exercised during installation to assure that all connections have been assembled liquid tight. Cold-weather installation (60°F) in particular usually requires additional heating to ensure that the adhesives joining the piping cure properly. FRP pipe has a UL listing for use in underground petroleum product service.

Double-Walled Piping

The application of double-walled pipe to underground storage systems is a recent innovation. While providing increased protection, double-walled piping is not universally available and requires a high degree of expertise to field-assemble. UL listing of this system has not yet been achieved.

TYPES OF PUMPING SYSTEMS

The two most common types of pumping systems are remote pumping systems and suction pumping systems. The remote system, illustrated in Figure 3, employs a submerged transfer pump which pumps liquid to the dispenser under positive pressure. Over the last several years, the remote system has gained increased acceptance due to a higher energy efficiency rating than suction pumps, as well as the advantage of needing only a single pump to service several dispensers. However, the remote system is unable to differentiate between normal operation (i.e., fuel dispensing) and leakage in the line. Application of line leak detectors will be discussed in Chapter 5.

The suction system employs negative pressure (suction) to pull the

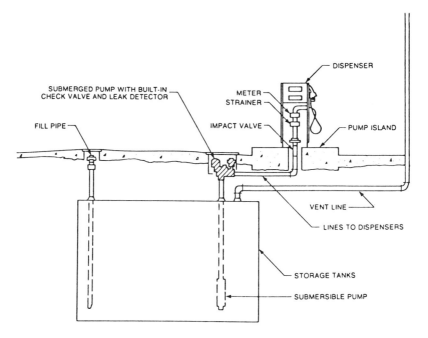

Figure 3. Schematic diagram of a remote pumping system. (Source: ANSI/API MPMS 6.3—1985.)[9]

liquid contents of a storage tank to a dispenser. Figure 4 is a schematic of a suction pumping system. A failure in the integrity of a piping system under negative pressure will result in readily identifiable signs of a problem (e.g., loss of pump prime, or air mixing with the liquid being dispensed). These signs are not readily apparent with a remote pumping system. An effective suction system typically employs one check valve located directly beneath the dispenser.

TYPES OF REGULATED SUBSTANCES

The 1984 Resource Conservation and Recovery Act Amendments stipulate under Subtitle I, "Regulation of Underground Storage Tanks," that certain materials stored underground are to be regulated. The federal statutory language defines a hazardous substance as "any substance defined in section 101(14) of the Comprehensive Environmental Response, Compensation and Liability Act of 1980 (but not including any substance regulated as a hazardous waste un-

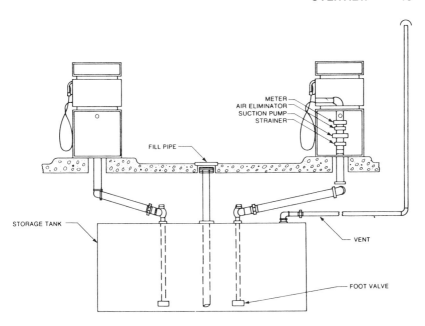

Figure 4. Schematic diagram of a suction pumping system. (Source: ANSI/API MPMS 6.3—1985.)[9]

der Subtitle C) and petroleum including crude oil or any fraction thereof which is liquid at standard conditions of temperature and pressure (60 degrees Fahrenheit and 14.7 pounds per square inch absolute)." Certain classes of tanks, however, are exempted from regulation even though they store "regulated substances." The exempt tanks include farm and residential tanks less than 1,100 gallons, heating oil tanks for consumptive use, septic tanks, stormwater and wastewater collection systems, and pipelines already regulated by the U.S. Department of Transportation. State regulatory programs, many of which are more stringent than the federal guidelines, may not exempt these materials. A complete list of all compounds regulated under Subtitle I can be found in Appendix A.

TYPES OF RELEASE DETECTION AND MONITORING SYSTEMS

An effective underground storage system program must integrate not only a preventive component, but detection capabilities as well. A diverse system that incorporates several different detection and monitoring strategies is necessary to address all aspects of a potential loss from an underground storage system.

The following chapters discuss the many different techniques available for leak detection and monitoring in an underground storage system. The different detection and monitoring techniques can be broken down into four broad categories: inventory monitoring, external storage system monitoring (including containment monitoring), internal storage system release detection, and piping release monitoring. A separate chapter is devoted to each of these subjects.

SUMMARY

Within the last several years, underground storage system design and construction has experienced more significant changes than in the previous 30 years. Driven by a heightened environmental awareness of the problem of leaking underground storage tanks and increasing regulatory pressures, new materials, innovative designs, and improved construction techniques have been developed to reduce the number and severity of leaks.

Leaks in underground storage systems are most commonly the result of corrosion, improper installation, or a lack of system maintenance. The types of materials that provide a high level of environmental protection and address the causes of leaks include cathodically protected steel, noncorrosive materials (e.g., fiberglass-reinforced plastic), and steel clad with corrosion-resistant materials. These materials can be incorporated into single- or double-walled tank and piping designs.

The substances stored in underground storage systems of environmental concern that are receiving regulatory attention include all the substances listed in section 101(14) of the Comprehensive Environmental Response, Compensation and Liability Act of 1980, and petroleum products. Storage systems containing petroleum (estimates of their number range from 2 to 6 million) far outnumber the other hazardous substance storage systems.

In developing a comprehensive system for detecting and monitoring leaks, it must be acknowledged that no one system will meet all the criteria of a fail-safe system in a comprehensive and cost-effective manner.

REFERENCES

1. Underwriters Laboratories, Inc., "Steel Underground Tanks for Flammable and Combustible Liquids" (UL 58). Northbrook, IL: Underwriters Laboratories, Inc., October 1976.
2. American Society of Mechanical Engineers, ASME Pressure Vessel Code, Section VIII, Division I, Boiler and Pressure Vessel Code. New York: American Society of Mechanical Engineers.
3. Garrity, Kevin C., "Cathodic Protection of Underground Storage Tanks" (HC-65). Medina, OH: Harco Corporation, January 1986.
4. American Petroleum Institute, "Cathodic Protection of Underground Petroleum Storage Tanks and Piping Systems," first edition (Publication 1632). Washington, DC: American Petroleum Institute, 1983.
5. Underwriters Laboratories, Inc., "Standard for Glass-Fiber-Reinforced Plastic Underground Storage Tanks" (UL 1316). Northbrook, IL: Underwriters Laboratories, Inc., July 1983.
6. Underwriters Laboratories of Canada, Inc., "Standard for Reinforced Plastic Underground Tanks for Petroleum Storage" (ULC-S615-1977). Scarborough, ON, Canada: Underwriters Laboratories of Canada, Inc., March 1977.
7. Division of Water, New York State Department of Environmental Conservation, "Recommended Practices for Underground Storage of Petroleum." Albany, NY: New York State Department of Environmental Conservation, May 1984.
8. Petroleum Equipment Institute, "Recommended Practices for Installation of Underground Liquid Storage Systems" (RP 100-86). Tulsa, OK: Petroleum Equipment Institute, 1986.
9. American Petroleum Institute, *Manual of Petroleum Measurement Standards* (first edition), Chapter 6 ("Metering Assemblies"), Section 3 ("Service Station Dispensing Metering Systems") (ANSI/API MPMS 6.3). Washington, DC: American Petroleum Institute, October 1983.

CHAPTER 2

Inventory Monitoring

Todd G. Schwendeman

CHAPTER CONTENTS

Introduction 21
Regulatory Requirements 23
Factors Affecting Inventory Monitoring 25
    Temperature 25
    Meter Accuracy 27
    Evaporation 29
    Gauging Accuracy 30
    Tank Geometry 31
Inventory Monitoring Techniques 32
    Manual Reconciliation Techniques 32
    Statistical Reconciliation Techniques 42
    Automatic Gauging Systems 47
Summary 48

Inventory Monitoring

Todd G. Schwendeman

INTRODUCTION

Inventory monitoring is a standard managerial practice for the successful operation of any type of business in which a commodity is being handled and distributed. Any material can be subject to losses when handled in quantity. In particular, monitoring the contents of an underground storage system requires a high degree of attention to account for a number of liquid handling and storage variables. Effective management of the liquid contents of an underground storage system will result in a safe environment for the system operator, employees, customers and neighbors; reduced air and water pollution; and increased operating profits. To date, the practice of underground storage system inventory monitoring as a leak detection technique has been primarily limited to motor fuels stored at retail distribution facilities.

The purpose of this chapter is to review the different methods of inventory monitoring and to describe the limitations and benefits of implementing some form of monitoring program. The existing regulatory requirements that mandate inventory monitoring and the various factors that affect the accuracy of the techniques will also be examined. The information presented in this chapter is designed to help the reader better understand what underground storage system inventory monitoring is, how it can be applied, and how the results can be used to detect leaks early.

Inventory monitoring of underground storage systems, when properly and conscientiously applied, is the most readily implemented and cost-effective technique for early leak screening. However, inventory monitoring is only one step in any leak detection program. Reliance upon inventory monitoring as the sole leak detection and monitoring technique will not achieve the sensitivity necessary to detect all types of releases from an underground storage system.[1] For this reason it should be used in conjunction with other, more sophisticated detection and monitoring methodologies.

The effectiveness of any inventory monitoring program relies principally on the degree of operator training and the conscientiousness of the individuals involved in the program. Data developed from EPA studies of underground storage system releases have indicated that only 15% of 12,500 documented release incidents were detected as a result of inventory monitoring programs.[2] Furthermore, results from the EPA Office of Toxic Substances study "Underground Motor Fuel Storage Tank National Survey" indicated that "it is very difficult to obtain accurate and usable inventory data."[3] The difficulty in obtaining high-quality inventory monitoring information has also been verified in other studies.[4] Improvement in the accuracy and reliability of inventory monitoring as an early leak screening technique will require expanded inventory monitoring training programs and improved data collection techniques. Nonetheless, recent insurance industry information has demonstrated that, because of an increased awareness of the leak detection benefits derived from an inventory monitoring program, an increased percentage of leaks have been identified by inventory monitoring.[19]

Inventory monitoring can be described as any means of recording and reconciling bulk storage transactions to assist a storage system operator in reducing bulk stock losses to achieve a high level of safety and pollution control while also maximizing profits. The importance of maintaining an inventory monitoring program is emphasized by the American Petroleum Institute (API) in Publication 1621: "The necessity of establishing and using an adequate accounting system . . . can not be overstressed. Even a very simple system may be adequate to control inventory of bulk liquid stock, but it is of utmost importance that it be done every day."[2]

The minimum data for the simplest inventory monitoring system should include:

- storage system identification number
- type of substance being stored
- all bulk liquid receipts
- all bulk liquid sales or use
- daily measurement of liquid stored

All of these data should be clearly recorded for each storage system at least once for each day that it is in operation. For storage systems that are out of service or that receive only intermittent use, a periodic examination of the inventory can provide useful information on the status of nonactive systems. Once collected, these data need to be analyzed and interpreted regularly to determine if unaccountable losses are occurring. A complete overview of underground storage system inventory monitoring as described in this chapter includes a review of existing and proposed regulatory requirements, a description of the various factors that affect the accuracy of inventory monitoring, and an overview of the different inventory monitoring techniques that are being applied to underground storage systems.

REGULATORY REQUIREMENTS

On November 8, 1984, the Resource Conservation and Recovery Act Amendments, containing Subtitle I, "Regulation of Underground Storage Tanks," were signed into law. The language of Subtitle I, Section 9003, directs the EPA to develop underground storage system leak detection requirements. These requirements are to include "an inventory control system together with tank testing or a comparable system or method designed to identify releases in a manner consistent with the protection of human health and the environment."[3] The federal underground storage regulatory program is currently being developed; proposed rules were to be issued in March 1987, with finalization anticipated in the spring of 1988.

Inventory monitoring requirements are also contained in the national fire codes. National Fire Protection Association (NFPA) Code 30, Flammable and Combustible Liquids Code, requires in Section 2-10 that "accurate inventory records or a leak detection program shall be maintained for all Class I liquid storage tanks for indication of possible leakage from the tanks or associated piping."[4] A Class I liquid is defined as a flammable liquid having a flash point below 100°F and a vapor pressure not exceeding 40 psi at 100°F. Nearly all states in the eastern two-thirds of the United States have adopted some edition of Code 30 as state law. Inventory monitoring is addressed in greater detail in NFPA 30A and NFPA 329. The Uniform Fire Code (UFC) applicable in the remaining western one-third of the United States also requires maintenance of underground storage system inventory records.[7,8] Table 1 summarizes the applicable state inventory control fire code requirements.

An additional organization of the federal government, the Occupational Safety and Health Administration (OSHA), also requires, in Section 1910.106(g), that all Class I liquid underground storage tanks maintain accurate inventory records.[9]

In addition to the requirements of NFPA, UFC, or OSHA, a num-

Table 1. State Inventory Monitoring Regulatory Requirements

State	Requirement	State	Requirement
Alabama		Missouri	
Alaska	UFC	Montana	UFC
Arizona	UFC	Nebraska	X[a]
Arkansas		Nevada	UFC
California	X	New Hampshire	X
Colorado	X	New Jersey	X
Connecticut	X	New Mexico	
Delaware	X	New York	X
Dist. of Col.	X	North Carolina	X[a]
Florida	X	North Dakota	
Georgia	NFPA	Ohio	X
Hawaii		Oklahoma	NFPA
Idaho	UFC	Oregon	
Illinois		Pennsylvania	X
Indiana	NFPA	Rhode Island	X
Iowa	X[a]	South Carolina	X
Kansas	X	South Dakota	UFC
Kentucky	NFPA	Tennessee	NFPA
Louisiana	NFPA	Texas	
Maine	X	Utah	UFC
Maryland	X	Vermont	NFPA
Massachusetts	X	Virginia	
Michigan	NFPA	Washington	UFC/NFPA
Minnesota	NFPA	West Virginia	NFPA
Mississippi		Wisconsin	X
		Wyoming	

[a]Regulations proposed as of July 1986.
X = no requirement.
NFPA = National Fire Protection Association.
UFC = Uniform Fire Code.

ber of states have developed other specific procedures. Various state approaches to inventory monitoring will be briefly described later in this chapter.

Even though all underground storage systems containing Class I fuels are currently subject to inventory reconciliation requirements, strict compliance by owner/operators is the exception rather than the rule. In addition, inventory monitoring is not employed to any significant extent outside of those underground systems storing petroleum products for retail distribution. Only in isolated circumstances have inventory monitoring provisions been strictly enforced by a designated regulatory authority. An increased emphasis on enforcement of existing state and local underground storage inventory monitoring

FACTORS AFFECTING INVENTORY MONITORING

Properly conducted inventory control procedures have been shown to be an effective first step for indicating the presence of an underground storage system leak.[4,5] Nonetheless, losses or apparent losses of inventory may occur in the course of daily operations. A number of factors can have significant impacts on apparent shrinkage or growth of inventory and contribute to these unavoidable losses: temperature variation, meter accuracy, evaporation, gauging accuracy, and tank geometry.[10] Depending upon the material being stored, all of these factors can affect the accuracy of inventory monitoring for liquids stored underground.

Temperature

The impact of temperature changes on inventory can be either a positive or a negative systematic effect, depending on the direction of temperature change and the magnitude of the coefficient of expansion of the stored substance. Temperature will influence the volume of the stored substance by two principal mechanisms: (1) temperature changes as a result of mixing delivered liquid with the liquid in the storage system, and (2) daily temperature variations in the storage system as a result of seasonal warming and cooling trends. Research results have demonstrated that motor fuel temperature changes following delivery can be upwards of 10°F, whereas day-to-day underground tank temperature fluctuations are typically less than 1°F.[11]

Although significant changes in stored motor fuel temperatures occur from season to season (e.g., summer to winter) the daily temperature changes relevant to inventory monitoring are not significant. As long as the retention time of the stored material is on the order of 7 to 14 days, seasonal fluctuations in underground temperature will have no significant impact on inventory monitoring.[10]

However, temperature variations resulting from the temperature differential of delivered and stored liquids can have a significant impact on inventory monitoring. Apparent inventory growth will occur when cold liquid is mixed with a warmer stored liquid (as in winter); conversely, inventory shrinkage will occur when a warm liq-

26 UNDERGROUND STORAGE SYSTEMS

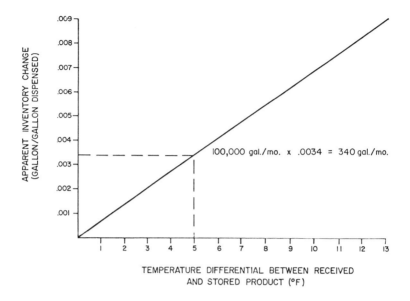

Figure 1. Temperature effects on inventory monitoring. (Source: Radian Corporation, "Analysis of Factors Affecting Service Station Inventory Control.")[10]

uid is mixed with a cooler stored material (as in summer). Although research results have shown that these effects cancel each other out over a long period of time, short-term inventory discrepancies can occur.[11]

The average volume temperature correction factor for gasoline at 60°F is 0.00068 per degree F. Figure 1 illustrates the apparent inventory changes attributable to temperature effects on gasoline temperature variations from 1°F to 13°F. For a storage system with a 100,000 gal/month throughput and an average 5°F temperature differential between delivered and stored product, an apparent inventory change of ±340 gal/month could occur as a direct result of temperature variations. The majority of inventory monitoring practices do not account for temperature-induced volume changes, although a limited number of independent retail motor fuel distributors employ temperature-compensated procedures for daily monitoring. Temperature-compensated inventory monitoring will be discussed in the final section of this chapter.

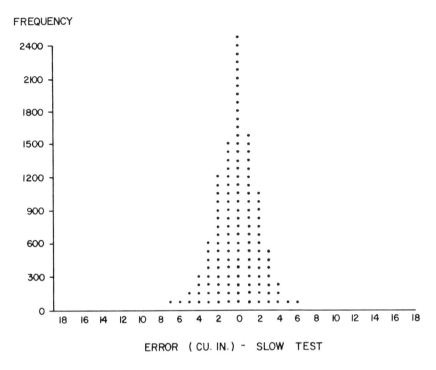

Figure 2. Histogram for slow flow meter testing. (Source: Radian Corporation.)

Meter Accuracy

It should be recognized that for storage systems using metered distribution of the stored substance, it is technologically impossible to calibrate dispensing meters to achieve and maintain 100% accuracy. The National Conference on Weights and Measures has established ±6 in./5-gal measure (0.5%)[3] as the legal tolerance level for motor fuel dispensing meters.

Meter accuracy effects on inventory monitoring can be either positive or negative. Results of meter accuracy studies indicate a small systematic effect of approximately −0.0002 gallons per gallon dispensed.[10] This statistic, however, relates only to the average metering error and not the range of errors that can be encountered. Figures 2 and 3 present histograms for slow and fast meter accuracy testing. Both distributions are similar, with the results varying around zero.

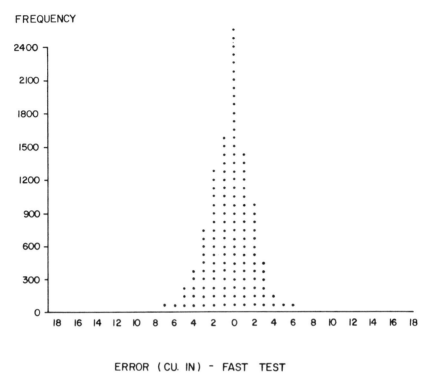

Figure 3. Histogram for fast flow meter testing. (Source: Radian Corporation.)

Note that meter inaccuracy for a given meter is not random but will give a consistent bias for loss or gain of the liquid stored.

As with temperature, the impact of meter accuracy on inventory monitoring is proportional to throughput (the greater the volume dispensed, the greater the error). Figure 4 presents the relationship between meter error in cubic inches in a 5-gal test and meter error in gallons per gallon dispensed.

Since the accuracy of inventory monitoring is a direct function of the quality of data obtained (stick and meter readings), an improperly calibrated or out-of-tolerance meter has the potential to mask or falsely indicate a leak. The impacts of meter inaccuracies can be reduced by periodically (e.g., monthly) checking the calibration of a storage system meter. A procedure for testing the accuracy of a gasoline dispensing meter as developed by API is presented in Appendix C.

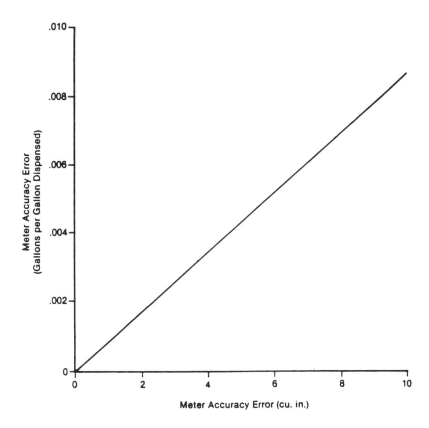

Figure 4. Relationship between meter error in gal per gal dispensed and meter error in in.³ in a 5-gal test. (Source: Radian Corporation.)

Evaporation

The impact of evaporative losses on inventory monitoring is a direct function of the volatility of the stored substance, the temperature at which it is stored, and the degree of vapor control exercised. Evaporative loss is principally due to vapor release during underground storage system filling and during routine storage system breathing. The total effect for motor fuel is approximately 0.0012 gal/gal throughput without any form of vapor control. About 85% of the vapor loss impacts are the result of losses during underground storage system filling and about 15% are storage system breathing losses. Vapor control at retail motor fuel dispensing facilities has demon-

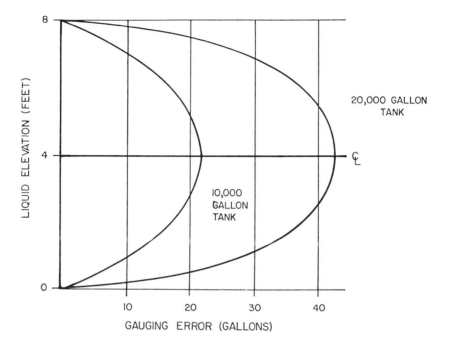

Figure 5. Error due to 1/8 in. gauging stick measurement in a 10,000- and 20,000-gal horizontal tank (8 ft 0 in. diameter).

strated that the losses can be reduced to as low as 0.0007 gal per gallon dispensed.[10]

Gauging Accuracy

The effects of gauging accuracy on inventory monitoring can be significant. A research study investigating the accuracy of gauge stick readings — the most common technique for gauging the contents of an underground storage system — indicated that a standard error for a wood gauge stick is 0.31 in.[10] Improved accuracy for use of the gauge stick can be achieved by replicate measurement and averaging of the results. (If s is the standard error for one measurement, the standard error of the mean of n independent measurements is s/n.) The greatest impact on inventory monitoring when using a gauge stick occurs when the liquid level is at the point of maximum diameter of the tank (e.g., midpoint). Figure 5 illustrates how gauging accuracy can vary in a given tank depending on the height of liquid stored.

Table 2. Typical Underground Steel Tank Configuration[a]

Gallons Capacity		Diameter (inch)	Length (inch)
Nominal	Actual		
560	564	48	72
1,000	1,034	48	132
2,000	2,005	64	144
3,000	3,008	64	216
4,000	4,018	72	228
5,000	5,076	72	288
6,000	6,132	72	348
7,000	6,011	96	192
8,000	7,890	96	252
10,000	10,152	96	324
10,000	9,994	108	252
12,000	12,032	96	384

[a]Source: Steel Tank Institute.

The accuracy of automatic gauging (the use of electronic liquid level measuring devices) is directly related to the adequacy of tank calibration data. However, as a result of the number of readings that can be attained from electronic systems and elimination of some of the "human element" inherent with stick gauging automatic gauging has the potential to improve inventory monitoring accuracy. These systems will be described in greater detail later in this chapter.

Tank Geometry

Tank geometry will have a systematic effect in either the positive or negative direction. Storage system volume will be affected by the calibration chart for the tank and does not account for inaccuracies in tank dimensions and position in the ground (slope). Calibration charts are developed uniformly for the different size tanks and are not developed for individual tanks. The relationship between tank geometry and volume or size depends on the ratio of tank diameter to length and the level of product in the tank.

Finite standards for storage tank sizes and types do not currently exist in the United States. UL permits a 5% tolerance in the construction dimensions for underground storage tanks. Tank configurations for steel cylindrical tanks with flat ends and capacities ranging from 560 gal to 12,000 gal are presented in Table 2.

INVENTORY MONITORING TECHNIQUES

Inventory monitoring of underground storage systems can be divided into three broad categories: manual reconciliation techniques, statistical reconciliation techniques, and automatic gauging systems.[16] Within each of these categories are a number of different techniques for reconciling inventory monitoring data. These different techniques and their relative performance will be reviewed in this section.

Manual Reconciliation Techniques

Manual techniques include static reconciliation, metered reconciliation, and temperature-compensated reconciliation. These techniques are the most commonly used underground storage system inventory monitoring procedures.

The equipment required to conduct manual inventory monitoring is limited to a gauge stick, water and product detection pastes, a tank calibration chart, data entry forms, and a calculator. A gauge stick is a pole of varnished hardwood or other hydrocarbon-compatible material that ranges in length from 6 to 10 ft and is about 1 in. wide and 3/4 in. thick. The stick is typically calibrated in inches with 1/8-in. gradations.

The gauge stick is used in conjunction with a calibration chart furnished by the tank manufacturer. The chart shows the number of gallons in the tank for each inch on the gauge stick. Each chart is calculated for a specific tank of particular dimensions and capacity. Manual tank gauging procedures and water level gauging procedures are detailed in Appendix B. Water level gauging procedures simply involve coating the measuring end of the gauge stick with water-sensitive paste. When the gauge stick touches water, the paste will change color, clearly indicating the depth of the water. The volume of water is subtracted from the full gauging volume to obtain the actual amount of material contained in the storage system.

API Publication 1621 recommends that all storage systems at a facility be gauged daily (preferably at the close of each shift), all storage systems be checked for the presence of water, and all sales and use be recorded at the same time. In addition, storage systems should be gauged before and after each delivery to verify that the proper amount of material ordered has been delivered.

The format of the data entry forms will vary with individual or company preference, number of storage systems at a facility, the

INVENTORY MONITORING   33

At least weekly, review the past week's daily inventory records.

A	B		C		D		E		F	G
	Opening Physical						Inventory		Closing Physical	Over
Product	Inventory	−	Sales	+	Receipts	=	Balance	−	Inventory	(Short)
TOTAL										

F
CLOSING INVENTORY STICK READINGS

Tank	Feet/Inches	Product	Gallons
1	/		
2	/		
3	/		
4	/		
5	/		
6	/		
7	/		
8	/		

Initialed by _____

Date _____

NOTE: FORWARD CLOSING PHYSICAL INVENTORY F TO OPENING PHYSICAL INVENTORY B ON NEXT DAY'S RECORD.

Figure 6. Sample daily inventory data entry form. (Source: API Publication 1621.)

physical layout of the storage systems at a facility, and the physical configuration of the storage system itself. Figures 6, 7, and 8 present different sample inventory data entry formats.

Recording inventory data is only the first step in employing inventory monitoring as a leak detection tool. The key step involves reconciling the inventory data daily. Without reconciliation, the operator will not obtain a true picture of how the storage system is operating.

Static reconciliation. For a variety of reasons, many underground storage systems may lack a metering system. Often these storage systems service the end user of the product. For those underground

MANIFOLDED TANK SYSTEM RECORDING SHEET

MANIFOLDED TANK SYSTEM I.D. NUMBER _____   TYPE OF FUEL _____

DAY	DATE	PHYSICAL INVENTORY MEASUREMENTS	TANK 1	TANK 2	TANK 3	TANK 4	TANK 5	TANK 6	TANK 7	TANK 8	LINE F* TOTALS
1		A. Opening stick (gals.) (Yesterday's line E)									
		B. Deliveries (gals.)									
		C. Total of fuel in tank (A+B)									
		D. Closing stick (inches)									
		E. Closing stick (gals.)									
		F. Fuel gone from tank (gals.) (C−E)									
2		A. Opening stick (gals.) (Yesterday's line E)									
		B. Deliveries (gals.)									
		C. Total of fuel in tank (A+B)									
		D. Closing stick (inches)									
		E. Closing stick (gals.)									
		F. Fuel gone from tank (gals.) (C−E)									
3		A. Opening stick (gals.) (Yesterday's line E)									
		B. Deliveries (gals.)									
		C. Total of fuel in tank (A+B)									
		D. Closing stick (inches)									
		E. Closing stick (gals.)									
		F. Fuel gone from tank (gals.) (C−E)									
4		A. Opening stick (gals.) (Yesterday's line E)									
		B. Deliveries (gals.)									
		C. Total of fuel in tank (A+B)									
		D. Closing stick (inches)									
		E. Closing stick (gals.)									
		F. Fuel gone from tank (gals.) (C−E)									

*Transfer Line E totals to the correct Inventory Review Sheet for this tank system.

Figure 7. Sample daily inventory data entry form. (Source: EPA, Chemical Advisory.)

INVENTORY MONITORING   35

INVENTORY REVIEW SHEET FOR TANKS WITH METERED DISPENSING PUMPS

TANK NO. _____   DISPENSING METER NO(S). _____   PROVING CAN _____
                       OR, TANK SYSTEM NO(S). _____   METER CHECK
                       IF MANIFOLDED TANKS _____

PART A

DIPSTICK INVENTORY

Column 1	Column 2	Column 3	Column 4	Column 5	Column 6	Column 7	Column 8	Column 9	Column 10
Date	Opening Dipstick Inventory (gallons)	Deliveries (gallons)	Total (2 plus 3)	Closing Dipstick Inventory (inches)	Closing Dipstick Inventory (gallons)	Gone from tank (4 minus 8)	Meter Sales (gallons) (from meter sheet)	Column 8 less than (-) or greater than (+) 7	7 & 8 : Subtract smaller from larger

Figure 8. Sample daily inventory data entry form. (Source: EPA, Chemical Advisory.)

storage systems which do not dispense the system contents through a meter, inventory monitoring can still be conducted by means of static reconciliation. This technique is based solely upon gauge stick readings at the beginning and end of a quiescent period during which the storage system is not in service. All inputs into the system should be gauged before and after the delivery.

The procedure for static reconciliation is relatively simple. The liquid level in the storage system should be accurately gauged immedi-

Date	Gauge Stick Reading (inches)		Volume (gallons)		Loss or Gain	Cumulative Total
	Opening	Closing	Opening	Closing		

Figure 9. Static reconciliation data entry form.

ately after the system has been taken out of service and recorded on a data entry form as illustrated in Figure 9. Before the system is placed back into operation, the liquid level should again be measured and recorded, before any withdrawals have occurred. The difference between the two entries will indicate the amount of loss or gain the system has experienced. This loss or gain figure should be carried forward for each period of storage system inactivity and a cumulative variance maintained by adding the gain or subtracting the loss from the previous entry.

An EPA-developed reconciliation technique compares the cumulative total of overages and underages to a series of motor fuel "action numbers." These statistically derived numbers will assist an owner, manager or operator in making a determination of what normal variances can be expected from operating an underground storage system. If during monthly reconciliation of the inventory data the cumulative total of underages (or overages) exceeds the action number, an investigation of the storage system's operation should be conducted. Table 3 illustrates the EPA "action numbers" for static reconciliation.

Static reconciliation for underground storage systems is directly affected by a number of variables. The degree of accuracy achievable by static reconciliation is directly proportional to the significance of the various variables (e.g., temperature, evaporation, or gauging accuracy). A higher degree of accuracy could be achieved when static reconciliation is applied to a material that has a very low coefficient of thermal expansion and a definitive color (such as used oil) than when it is used for motor fuels or most solvents. The size of the storage system and the frequency with which data are recorded will also affect static reconciliation accuracy. A smaller system which is gauged daily will experience a higher degree of leak detection accuracy than a larger system which is gauged infrequently.[16]

Table 3. Action Numbers for Static Reconciliation[a]

Inactive Period	Action Number (gallons)
1	14
2	20
3	25
4	29
5	32
6	35
7	38
8	40
9	43
10	45
11	47
12	49

[a]Source: EPA, *More About Leaking Underground Storage Tanks*.[1]

In essence, static reconciliation has the ability to detect volume changes that appear as consistent and increasing negative (or positive due to water infiltration) trends. Through conscientious observation of the daily underages or overages, a trend indicating a loss or gain can be observed. This phenomenon typically manifests itself by a steady increase in the size of the underage or overage. Leak detection accuracy will be a function of tank size (greater accuracy can be achieved on smaller tanks). This technique, as with all other forms of inventory monitoring, is a leak screening technique that should be used in conjunction with more sophisticated leak detection and monitoring systems.

Metered reconciliation. Underground storage systems that are used to store liquids for resale commonly are equipped with metered dispensers. At such facilities a daily reconciliation should be conducted, using the following data: meter readings, delivery volumes, unmetered on-premises use, and an opening (or closing) gauge stick reading. Simple mathematics will supply the information required to determine the daily inventory balance:

Opening (or Closing) Gauge Stick Inventory + Deliveries
− Sales − Unmetered Use = Book Inventory

Metered reconciliation is completed by comparing the measured or opening inventory with the book or calculated inventory. Small variations between these two numbers are common because of certain unavoidable errors in measurement inherent in the process. Comparison of the opening inventory reading with the book inventory value will often reveal apparent system overages or underages. Analyses of these overages or underages are used to determine if the system can be assumed to be leak free or if there is an indication of a potential leak. Although theft and computation errors can be the cause of inventory variances, the owner/operator should be particularly sensitive to the possibility of leakage.

Inventory monitoring records from a storage system that does not have a leak should have daily overages and underages that fluctuate randomly around zero. A potential leak in the system may be possible when a certain "trigger" value is exceeded, a sudden large-volume loss occurs, or a pattern of continual but growing daily losses is apparent.

A trigger value is an inventory reconciliation reference point which, when exceeded, indicates a potential problem. A variety of manual inventory monitoring trigger values have been established. Trigger values based on monthly throughput range from the API-established 0.5% (150 gal for a tank with 30,000 gal/month throughput) to several state requirements that range from 0.1 to 1.0%. Many states have also specified a volume figure which, if exceeded, will require that an investigation be initiated. These values range from 50 to 250 gal for any given day or for several days in succession.

The EPA has developed a metered inventory reconciliation technique for motor fuels based on a statistically derived cumulative number of shortages that may occur during a specific time interval.[1] Comparison of the number of shortages for a storage system during 30-day periods with an "action number" is an easily applied method for determining the presence of a potential problem. Table 4 illustrates the "action numbers" for a full year of metered inventory reconciliation data.

Temperature-compensated reconciliation. Temperature-compensated reconciliation is designed to differentiate between a storage system leak and motor fuel shrinkage that occurs as a result of mixing warm delivered motor fuel with cooler in-tank fuel.[17] The application of this technique is limited to those situations in which temperature-adjusted material is delivered to a storage system.

Through field experimentation, a petroleum retailer has determined that a leak detection temperature correlation exists between the measured gain (or loss) of motor fuel and the calculated gain (or loss).[17] If a system is not experiencing any abnormal losses (leaks), then the measured change in volume of the motor fuel should fall

Table 4. Action Numbers for Metered Reconciliation[a]

30-Day Period	Action Number[2]
1st	20
2nd	37
3rd	54
4th	69
5th	25
6th	101
7th	117
8th	133
9th	149
10th	165
11th	180
12th	196
30-business-day inventory period	

[a]Source: EPA, *More About Leaking Underground Storage Tanks*.[1]

within a range of 30% to 70% of the calculated temperature-induced volume change. Any comparison of the measured vs calculated volume change that is outside of the 30% to 70% range is considered to indicate a potential leak.

The temperature-compensated reconciliation technique is best explained through examination of Table 5, which details a record of a retail motor fuel facility's monthly inventory monitoring summary. This record is a wintertime example, in which the volume of motor fuel purchased is temperature-corrected to the volume at 60°F. The difference between the amount that was ordered (net) and the amount that was delivered (gross) is in the Calculated Volume column. The actual gauge stick recorded difference between what was ordered and what was delivered is in the column labeled Measured Volume. To determine if the delivered motor fuel is within the 30% to 70% correlation, the measured volume change is divided by the calculated volume change and converted to a percentage.

In the winter example presented in Table 5, both the leaded and

Table 5. Example of Temperature-Compensated Inventory Monitoring[a]

Product	Gross (gallons)	Net (gallons)	Calculated Volume (gallons)	Measured Volume (gallons)	Measured (100) Calculated
Leaded	42,216	42,638	422	141	33%
Unleaded	83,675	84,239	564	368	65%
Super Unleaded	15,212	15,506	294	9	3%

[a]Source: Kocolene Oil Corporation.

unleaded motor fuel deliveries fell within the 30% to 70% correlation. The super unleaded fuel tank inventory was outside the correlation (3%) and warrants further leak investigation. In addition, the influence of water entering the storage system could have caused a correlation between net and gross gallons to have exceeded the 70% ceiling and thus also triggered an investigation.

Temperature-compensated inventory monitoring is applied on a monthly basis and enables an underground storage system owner/operator to refine an operation's leak detection capabilities. The technique has yet to be thoroughly investigated and scientifically validated, and it is limited to those substances that can be temperature-adjusted prior to delivery.

Manual inventory monitoring investigations. Studies have been conducted evaluating the various manual inventory monitoring reconciliation techniques.[4] An API study investigated the accuracy and reliability of 10 different reconciliation methods. Each method used a different trigger value to initiate an inventory monitoring investigation. The accuracy of each method was evaluated using inventory records of known leaking and nonleaking storage systems. A summary of the different methods evaluated in the study is presented in Table 6.

The results of the study indicated that only three methods had acceptable performance, defined as high leak detection accuracy and low false alarm tendency. Only one of the three acceptable reconciliation procedures, the API procedure (the Michigan procedure is similar to the API procedure), is a manual reconciliation method. All other manual techniques were either incapable of distinguishing between leaking and nonleaking storage systems (i.e., had high leak detection accuracies but also had very high false alarm tendencies), or incapable of detecting a significant number of leaking tanks (e.g., fewer than 50% detected). Table 7 summarizes the results of the performance evaluation.

For clarity, the study results can be organized into three distinct groups with regard to their performance:

Group 1: Procedures which can discriminate relatively well between leaking and nonleaking storage systems.

Group 2: Procedures which cannot effectively discriminate between leaking and nonleaking storage systems.

Group 3: Procedures which are poor at identifying the majority of leaking storage systems in the database.

Table 6. Summary of Procedures Selected for Evaluation[a]

Procedure Name	Time Period	Abnormal Loss Value(s) Which Signal Possible Leak — Value
API-1621	Monthly	Loss exceeds 0.5 percent of deliveries
EPA Action Numbers	Monthly	Number of days with losses exceeds a specific number (20 for month 1, 37 over 2 months, etc.)
New York State	10 Days	Loss or gain exceeds 0.75 percent of deliveries
Florida State	Weekly	Loss or gain exceeds 1.0 percent of deliveries
Michigan	Monthly	Same as API-1621
	Monthly	Number of days showing a loss exceeds 18 for any month
California	Daily	Loss or gain exceeds 100 gallons
	Weekly	Loss or gain excees 5 percent of sales or 100 gallons whichever is greater but should never exceed 350 gallons
	Monthly	Loss or gain exceeds 0.5 percent of sales or 100 gallons, whichever is greater
API Member Company	Daily	Loss or gain exceeds 300 gallons
	3 Days	Loss or gain exceeds 150 gallons per day
	Monthly	Loss or gain exceeds 0.5 percent of sales
Statistical Procedures X/Y	Monthly	Statistical evaluation. No specific imbalance value used.
Kocolene Oil Corporation	Monthly	Requires temperature-compensated delivery volumes. Measured gains or losses greater than 70% or less than 30% of the expected gain or loss (net vs. gross deliveries).

[a]Source: API, "Analysis of Existing and Proposed Underground Storage Tank Inventory Control Procedures."[4]

Table 8 places each procedure into one of these three performance groups and ranks each procedure by decreasing accuracy within each group. The results of the statistical procedures evaluation will be discussed later in this section.

From these study results, it appears that a manual inventory reconciliation imbalance that exceeds 0.5% of monthly throughput should

Table 7. Comparison of the Performance of Evaluated Inventory Monitoring Procedures[a]

Procedure	Leak Detection Accuracy — Percent of Leaking Tanks Correctly Identified	False Alarm Tendency — Percent of Nonleaking Tanks Incorrectly Identified as Leaking
API-1621	100	29
Michigan	100	29
New York	100	83
California	100	54
API Member Company	94	62
Statistical Procedure X	75[c]	32[c]
Statistical Procedure Y[b]	63[c]	27[c]
Florida	31	5
EPA Action Numbers	18	0
Kocolene Oil Corporation	d	d

[a]Source: API, "Analysis of Existing and Proposed Underground Storage Tank Inventory Control Procedures."[4]
[b]Statistical Procedure Y identified a number of tanks as "too close to call" in terms of identifying the leak status. In reporting the results here a conservative approach was taken and those tanks identified as "too close to call" were identified as possible leaking tanks.
[c]The values reported here can vary depending on how the results from these procedures are interpreted.
[d]Not evaluated because of a lack of temperature-compensated delivery data.

be investigated. In addition, this study concluded that use of a specific volume number trigger value (e.g., 50 gal, 250 gal) is less effective in detecting leaks over the broad range of throughput volumes and will be effective only in a catastrophic loss situation.

Statistical Reconciliation Techniques

Computer-based statistical reconciliation programs are being applied to inventory monitoring of underground storage systems with increasing frequency. Statistical reconciliation uses the same basic inventory monitoring data (e.g., gauge stick readings and metered withdrawals) as manual reconciliation. Developers of the statistical techniques claim that the programs are capable of detecting leaks as small as 1 to 2 gal/day.

These programs use statistical techniques to isolate trends in daily inventory records. The procedures not only identify the presence of an indicated loss, but attempt to attribute the variation to a specific

Table 8. Comparison of Evaluated Inventory Monitoring Procedures by Performance Group[a]

GROUP 1: Procedures which can discriminate relatively well between leaking and nonleaking tanks.

Procedure	Leak Detection Accuracy	False Alarm Tendency
API-1621	100	29
Michigan	100	29
Statistical Procedure X	75	32
Statistical Procedure Y	63	27

GROUP 2: Procedures which cannot effectively discriminate between leaking and nonleaking tanks.

Procedure	Leak Detection Accuracy	False Alarm Tendency
New York	100	83
California	100	54
API Member Company	94	62

GROUP 3: Procedures which cannot effectively identify leaking tanks.

Procedure	Leak Detection Accuracy	False Alarm Tendency
Florida	31	5
EPA Action Number	18	0

[a]Source: API, "Analysis of Existing and Proposed Underground Storage Tank Inventory Control Procedures."[4]

cause. Inventory monitoring record discrepancies that can be identified include record-keeping variations, dispensing meter variations, temperature effects, vapor losses, unexplained additions and removals, gauge stick variations, and storage system leaks. The statistical procedures attempt to identify the causes of the variations by assuming that each major cause of a discrepancy exhibits a distinct and consistent pattern throughout the records.

To achieve the maximum benefit from a statistical analysis of a storage system's inventory monitoring records, the following inputs should be submitted:

- at least 20 consecutive days of inventory records
- gallon amount and calendar dates and times for deliveries, pump totals, and gauge stick readings
- gallon amounts specified as net or gross
- basic storage system information: capacity, type, material stored, pumping system, and type of vapor recovery devices

- general facility information: locale (city and state), days and hours of operation

Figures 10 and 11 are sample statistical inventory review forms. Figure 10 is a "Daily Inventory" form and is the minimum information needed. Figure 11, a "Thermal Shrinkage" form, is additional information which can be used but is not mandatory.

The outputs obtained from a statistical analysis can include estimates of:

- storage system leakage rate
- storage system inflow or outflow
- dispensing meter inaccuracy
- conversion chart inaccuracy
- vapor loss
- temperature expansion or contraction
- delivery variations
- gauge stick variations
- one-time unexplained losses

The ability to obtain additional information on the operation of a storage system may be of significant value to an owner or operator and may improve the accuracy of an operator's inventory monitoring procedures. Statistical reconciliation has an additional benefit of eliminating the human interpretive element that plays a major role in manual inventory monitoring techniques. An operator does not have to make a judgment on the existence of a leak based on a personal interpretation of the inventory reconciliation data. Statistical analysis of inventory records has been recommended by Environment Canada and Canadian provincial and territorial governments[18] and is an option in the State of Maine Underground Storage Tank Regulations for investigating a suspected release before precision tank testing is ordered.

The inventory procedures performance evaluation conducted by API demonstrated that statistical techniques have the ability to detect leaks from analysis of inventory data. The results are detailed in Table 7. The statistical reconciliation techniques that were analyzed were ranked in Group 1—"procedures which can discriminate relatively well between leaking and nonleaking storage systems"[4] (Table 8).

Statistical reconciliation is typically used once or twice per year by a

TANK SURVEY REPORT—DAILY INVENTORY

Submit forms to:

Station No._____
Address_____
City_____ State_____
Brand_____

For. day _____

Tank	[A] Product sold by tank (gal)	[B] Product delivery (gal)	[C] Product inventory by dipstick		[D] Tank Water inventory by dipstick	
			Dipstick level	Product (gal)	Dipstick of water	Water (gal)
Morning	____am	____am	____am		____am	
1	____	____	____	____	____	____
2	____	____	____	____	____	____
3	____	____	____	____	____	____
4	____	____	____	____	____	____
5	____	____	____	____	____	____
6	____	____	____	____	____	____
Evening	____pm	____pm	____pm		____pm	
1	____	____	____	____	____	____
2	____	____	____	____	____	____
3	____	____	____	____	____	____
4	____	____	____	____	____	____
5	____	____	____	____	____	____
6	____	____	____	____	____	____

[A] Total (sum) of all pump readings from dispensers accessing this tank.
[B] Total deliveries made today to this tank (enter 0 if no delivery was made).
[C] Enter dipstick level (inches) and the corresponding product (gallons) from the manufacturer's dipstick calibration chart.
[D] Enter dipstick level of tank water (inches) and the corresponding water volume (gallons) from the manufacturer's dipstick calibration chart.

Date_____
Recorded by_____ A-29
Affiliation_____

Figure 10. Sample statistical inventory review form—daily inventory. (Source: Entropy Ltd.)

46 UNDERGROUND STORAGE SYSTEMS

TANK SURVEY REPORT—THERMAL SHRINKAGE

Submit forms to:

Station No. _____
Address _____

City   State
Brand _____

For day _____

	[A] Before delivery: product inventory by dipstick			[B] After delivery: product inventory by dipstick			[C] Dipstick computed product delivery
	Dipstick level (inch)	Product (gal)	Midlevel temp. (°F)	Dipstick level (inch)	Product (gal)	Midlevel temp. (°F)	(gal)
Morning	____ am			____ am			____ air temp.
1							
2							
3							
4							
5							
6							
Evening	____ pm			____ pm			____ air temp.
1							
2							
3							
4							
5							
6							

[A] Immediately before delivery, enter dipstick level (inches), the corresponding product (gallons) from the manufacturer's dipstick curve and the product temperature (°F) at half the dipstick level. Enter time (hours and minutes).

[B] Enter same information as for [A] taken immediately after delivery.

[C] Subtract product gallons under [A] from product gallons under [B]. Enter air temperature (°F) at delivery time using same thermometer.

Date _____
Recorded by _____ A-30
Affiliation _____

Figure 11. Sample statistical inventory review form—thermal shrinkage. (Source: Entropy Ltd.)

large company with many facilities. The cost to analyze inventory records for one month of data is $40 to $80 per storage system.[14]

Automatic Gauging Systems

Within the last several years, automatic gauging systems for application to underground storage systems have been developed. These systems are installed inside a tank and provide continuous monitoring of liquid level changes. By means of a computer interface, they can convert level measurements to volume measurements and reconcile the inventory data.[14]

To assist an owner or manager in evaluating the many different types of automatic gauging systems currently available, a set of desirable characteristics needs to be established. These characteristics include:[21]

- ability to measure a variety of parameters (e.g., liquid level, water level and temperature) for each storage system
- sufficient precision and accuracy to detect a change in level comparable to other forms of measurement (gauge sticking, meters, tightness testing, etc.)
- a more sensitive detection of rate of volume change that operates when the storage system is inactive
- the capability of monitoring all storage systems present at a facility
- ease of installation in either new or existing underground storage tanks
- ease of interface with other automated systems at a storage facility
- a flexible sensing device to accommodate storage tanks with different diameters
- limited maintenance and ease of repair
- ability to measure liquid temperature changes and provide volume compensation
- a high-level alarm to prevent overfills

For an automatic gauging system to record inventory data accurately and operate as a leak monitoring system, physical characteristics of the storage system (e.g., tank geometry and slope) and of the material being stored (e.g., specific gravity and coefficient of thermal expansion) must be taken into consideration.

A variety of technologies have been applied to automatic liquid level measurement devices in underground storage systems. These technologies include:

capacitance	servo measurement
ultrasonics	plumb-bob measurement
resistivity	hall effect switches
magnetostriction	reed switches

Several systems have the advantage of employing a more sensitive rate of volume change mode while the storage system is inoperative. Since most of the automatic gauging systems temperature-compensate, a liquid level change during a period of inactivity may signal the loss of some of the stored material. An automatic gauging system is one of the few pieces of equipment that possess both inventory management and limited leak detection capabilities.

The liquid level sensing capabilities of these different systems will range from 0.7 in. to 0.001 in. Costs for these systems currently vary from $3,000 to $12,000 for a typical three-tank storage facility. Table 9 summarizes the principal performance characteristics of automatic gauging systems. The sensitivity of these systems requires care during handling and installation. In addition, the sophisticated instrumentation on which automatic gauges are based will need some form of periodic maintenance.

SUMMARY

Inventory monitoring is the most cost-effective and easily implemented form of leak monitoring available. However, attention must be devoted to training of personnel performing the tasks and to conscientious application of the monitoring technique. Training should include educating personnel on the consequences of a release and the techniques to identify the early warning signs of a release through inventory monitoring. Inventory monitoring is a screening technique that must be applied with other leak monitoring methodologies to achieve the sensitivity necessary to detect liquid losses from a storage system.

Regulations mandating inventory monitoring have been in effect for some time but have received only marginal enforcement. Inventory monitoring guidelines are specified in NFPA 30, UFC, OSHA (Section 1910.106[g]), and in all specific state underground storage regulatory programs.

The accuracy of inventory monitoring of underground storage systems will be affected by a number of factors, including temperature, meter accuracy, evaporation, gauging accuracy, and tank geometry.

INVENTORY MONITORING 49

Table 9. Performance Characteristics of Automatic Gauging Systems[a]

Type of Device	Installation	Liquid[b] Height Measurement Accuracy	Estimated Release Rate Detection Capability (gph)[c]	Measurement Precision (minimum detectable liquid level change)[b]	Approximate Cost[d]	Advantages	Disadvantages
Capacitance device	New or retrofit	±0.1 in. during operation of tank	0.1	±0.001 in. during quiet tank periods	$3,000–$7,500	Considerable field experience. Not affected by specific gravity changes or foaming problems.	Many require ground plane. Needs separate water sensor.
Ultrasonic device	New or retrofit	±0.1 in.	0.1	±0.001 in. during quiet tank periods	$3,400–$7,200	Can perform temperature and water level measurements without separate circuits	Has "Dead Band," 6 to 18 in. above or below transducer, which cannot be measured.
Resistance sensor device	New or retrofit	±0.2 in.	12	±0.125 in.	$5,500	Easily installed. Clear of obstruction.	Vent/equalizer must be kept.
Magnetostrictive device	New or retrofit	±0.035 in.	4	±0.035 in.	$9,000–$12,000	Considerable field experience	Expensive.
Servo-level gauge	New or retrofit	±.01 in.	1	±0.01 in.	$6,000–$8,000	Gives continuous level readings. Requires explosion-proof design for use in hydrocarbon tanks.	May be difficult to install in existing tank.

Table 9. Continued

Type of Device	Installation	Liquid[b] Height Measurement Accuracy	Estimated Release Rate Detection Capability (gph)[c]	Measurement Precision (minimum detectable liquid level change)[b]	Approximate Cost[d]	Advantages	Disadvantages
Plumb-bob device	New or retrofit	±.01 in.	1	±0.01 in.	$6,000–$8,000	Easy access for maintenance or installation. Requires explosion-proof design for use in hydrocarbon tanks.	No continuous data collection.
Hall effect switch device	New or retrofit	±0.003 in.	0.3	±0.003 in.	$4,000	Very sensitive.	Limited field testing.
Reed switch device	New or retrofit	±0.07 in.	7	±0.07 in.	$4,500	Field-serviceable.	Float could stick in some situations.

[a]Source: American Petroleum Institute, "Automatic Gauging Systems as Release Detection Techniques."
[b]These detection ranges are based primarily on manufacturer or distributor claims, which have not been verified through independent testing by API or its contractors. Reported detection ranges are at 60°F for capacitance, ultrasonic, and magnetostrictive devices; no temperature data available for the other devices.
[c]These estimates are based on an assumed 4-hr inactive tank measurement period and a required level change of three times the purported liquid-level-change measure precision.
[d]Cost estimates are for the equipment required to outfit a three-tank storage system.

Depending upon the type of substance stored, each of these factors will influence the leak detection capabilities of an inventory monitoring system.

Inventory monitoring can be conducted a number of different ways, depending upon the resources available. The three forms of inventory monitoring are manual reconciliation, statistical reconciliation, and automatic gauging. Liquid level readings and metering data (if the system is metered) form the base of inventory monitoring. Actual leak detection is obtained from reconciliation of the base data. Regardless of what form of inventory monitoring is applied to an active underground storage system, daily recording of data is essential.

REFERENCES

1. Office of Toxic Substances, U.S. Environmental Protection Agency, *More About Leaking Underground Storage Tanks: A Background Booklet for the Chemical Advisory*. Washington, DC: U.S. Environmental Protection Agency, October 1984.
2. U.S. Environmental Protection Agency, "State Release Incident Survey: Preliminary Draft." Washington, DC: U.S. Environmental Protection Agency, May 1986.
3. Office of Toxic Substances, U.S. Environmental Protection Agency, "Underground Motor Fuel Storage Tank National Survey." Washington, DC: U.S. Environmental Protection Agency, June 1986.
4. American Petroleum Institute, "Review and Analysis of Existing and Proposed Underground Storage Tank Inventory Control Procedures: Revised Draft Report." Washington, DC: American Petroleum Institute, May 1986.
5. American Petroleum Institute, "Recommended Practice for Bulk Liquid Stock Control at Retail Outlets," third edition (Publication 1621). Washington, DC: American Petroleum Institute, 1977.
6. Resource Conservation and Recovery Act Amendments, Title VI ("Underground Storage Tanks"), Subtitle I ("Regulation of Underground Storage Tanks"). *Congressional Record*, October 3, 1984, H11120–H11123.

CHAPTER 3

External Tank Release Detection and Monitoring

Todd G. Schwendeman

CHAPTER CONTENTS

Introduction	55
Contaminant Transport	57
Liquid Transport	57
Dissolved-Phase Transport	59
Vapor Transport	61
Release Detection Techniques	64
Groundwater Detection Techniques	64
Detection Wells	65
Soil Sampling	73
Dyes and Tracers	73
Surface Geophysics	81
Vadose Zone Vapor Detection Techniques	83
Grab Sampling of Soil Cores	83
Surface Flux Chambers	84
Downhole Flux Chamber	84
Accumulator Systems	86
Ground Probe Testing	88
Physical Inspection	90
Visual Inspection	90
Integrity Testing (Tank Shell)	90
Release Monitoring Systems	93
Observation Wells	94
Design	94
Construction	95
Installation	96
Sampling	97
Vapor Wells	103
Design	104
Construction/Installation	104
Sampling	104

54   **UNDERGROUND STORAGE SYSTEMS**

> U-Tubes 107
>   Design 108
>   Construction/Installation 109
>   Sampling 109
> Secondary Containment Monitoring 109
> Summary 111

External Tank Release Detection and Monitoring

Todd B. Schwendeman

INTRODUCTION

Of the many different forms of underground storage system release detection and monitoring, external tank techniques have the widest application and the greatest historical use. Even though many creative technologies and concepts are being applied to underground storage system detection and monitoring, such fundamental technologies as soil sampling and wells are utilized most frequently.

Before extensive discussion, it is useful to define several terms and make several significant distinctions. This chapter is specific to underground storage *tank* release detection and monitoring. The leak detection and monitoring approaches for a tank differ in logic and application from the approach for piping. Significant differences between the storage tank and the liquid distribution piping include exposed surface area, areal extent, operating pressures, design construction and installation standards, and ease of access and repair. Release detection and monitoring of buried liquid piping will be addressed in detail in Chapter 5.

Distinctions should also be made between release detection and release monitoring, even though the purpose of both technologies is to identify unconfined substances. Release detection is the physical act of confirming that a release has occurred. This act is usually performed at a given point in time (usually after a leak has been suspected), is a singular event, and is not a continuously operating system. Release monitoring (sometimes referred to as release effects monitoring), on the other hand, consists of a periodic program of sampling and analysis or observation, usually from a fixed monitoring system, that spans an extended period in time. A continuous monitoring system will normally function without interruption so that a release can be detected as early as possible. This system is commonly considered a part of an underground storage system's equipment and normal operation.

Location of an external release detection and monitoring system is critical. It should be readily apparent that the closer to the source monitoring is conducted the earlier the leak will be detected. A fundamental rule of liquid transport in the subsurface is that an unconfined liquid will seek the path of least resistance as it migrates away from the source of the release. Therefore, placement of an external monitoring system within the storage system excavation zone will provide greater probability for the detection of a release.

Release detection and monitoring system location is an important issue when installation of a system for new tanks or retrofit of existing tanks is considered. It is relatively simple to design, construct, and install a monitoring system for a *new* tank. Much of the information required for accurate and reliable monitoring data is well known: tank location, other buried subsurface structures, hydrogeologic conditions, etc. The expense for a monitoring system at a new installation is principally capital costs and can be amortized over the life of the facility.

The cost of operating and maintenance for most systems is not significant. Release monitoring at an *existing* location, however, is usually financed from the operating budget and is complicated by many unknown elements. A thorough investigation of the subsurface conditions and other influences must be completed before a monitoring system is selected and installed. If the subsurface conditions cannot be accurately quantified, the monitoring system may have to be installed at a greater distance from the tank, with a resultant decrease in early warning capabilities. Quality hydrogeologic information at an existing site will permit selection of the most appropriate, site-specific monitoring system, and will result in higher performance and increased operational confidence.

Chapter 3, "External Tank Release Detection and Monitoring," has six sections: Introduction, Contaminant Transport, Release Detection Techniques, Release Monitoring Systems, Secondary Containment Monitoring, and Summary. The introduction provides a brief overview of release monitoring and detection outside of the storage tank. The section on contaminant transport discusses liquid and vapor transport in the subsurface; unconfined substance subsurface transport characteristics will directly influence the type of technique or system that will be most effective. The third section, on release detection techniques, addresses a number of different methodologies, some simple and some sophisticated, to confirm that a release has occurred. The fourth section, on release monitoring systems, has been confined to the three most applicable external monitoring systems: observation wells, U-tubes, and vapor wells. Tank tightness testing, a widely practiced form of release detection (or confirmation), is discussed in detail in Chapter 4. The fifth section of the chapter describes release monitoring techniques applicable when a tank is provided with a secondary barrier, which creates an enclosed space around the tank. In the event of a release, this barrier prevents contact of the stored material with the environment.

CONTAMINANT TRANSPORT

Effective external detection and monitoring requires knowledge of how substances released from a storage system will transport in the subsurface. Different stored substances can be transported as liquids, as vapors, or as both. The manner in which an unconfined substance transports depends on the type of substance, the rate and volume of the release, and the surrounding hydrogeologic environment. This section is designed to present an overview of contaminant transport. The discussion has been limited to organic substances, as these are the principal materials of concern stored in the subsurface.

Liquid Transport

When an organic liquid is released from an underground storage system, the substance will tend to migrate vertically downward under the force of gravity, with some lateral spreading.[1] The movement of hydrocarbon at various rates and through stratified soil is illustrated in Figure 1. Downward vertical migration of a released substance will be interrupted by one of three events: (1) the substance will be adsorbed by the soil, (2) the substance will encounter an impermeable geologic formation, or (3) the substance will reach the water table.

In almost all release cases, some volume of the unconfined substance remains behind the migrating front. This residual material, adsorbed by the soil in the vadose (unsaturated) zone, can be the source of continual contamination by leaching via precipitation, infiltration, and groundwater fluctuations.[2,3] The technology of contaminated soil treatment is only beginning to be addressed. Research and development efforts regarding treatment techniques designed to remedy residual soil contamination should receive greater attention.

Immobilization of a released substance by soil adsorption depends on three factors: (1) size of the soil grains (e.g., texture—fine-grained

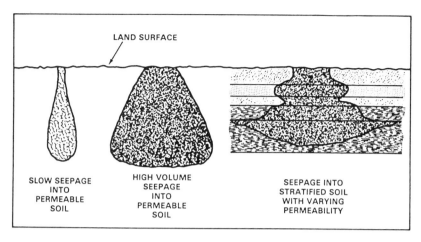

Figure 1. Transport of hydrocarbon in the subsurface. (Source: API Publication 1628.)[3]

vs coarse-grained), (2) percentage of organic material present, and (3) the individual characteristics of the liquid. The volume of porous geologic material that is required to immobilize a release of hydrocarbon, for example, can be grossly approximated from the following equation:[2]

$$\frac{\text{Soil required to attain}}{\text{immobile saturation (yd}^3\text{)}} = \frac{0.2V}{PS_r}$$

where V = volume of release (barrels)
P = soil porosity (percent)
S_r = residual hydrocarbon saturation

The following values typically can be used for residual saturation (S_r) when silts and clays are encountered: 0.10 for gasoline, 0.15 for diesel and light fuel oil, and 0.20 for lube and heavy oils. Unfortunately, for this equation to be an accurate estimate, homogenous geologic conditions must be present, a situation that rarely occurs. As a general rule, the finer the geologic material, the greater the adsorptive capacity for fugitive products.

As the body of an unconfined release migrates vertically downward, its path will be affected by the permeability of the geologic materials in which it is migrating. If the substance encounters an impermeable geologic layer, the direction of movement will be altered. The substance will spread laterally until it either becomes immobile, flows around the impermeable layer and continues downward migration, or outcrops on the surface.

Groundwater contact is the most serious consequence of subsurface migrating contaminants, because of the potential or actual usage of the groundwater as drinking water. Should contact with the groundwater occur (at the interface between the vadose zone and saturated zone), the unconfined substance will float on top of the water table (immiscible liquid) or dissolve into the groundwater (miscible liquid). Besides miscibility, other physical properties important to the migration of a leaking substance in the subsurface are viscosity, surface tension, evaporation rate, and vapor density.[4] Migration or subsurface transport of immiscible organic liquids is governed by a different set of factors from migration or subsurface transport of dissolved contaminants in the aquifer system, and is discussed later. In the following section, the factors affecting dissolved phase movement and migration are discussed.

Dissolved-Phase Transport

There are four primary physical principles that regulate the migration of dissolved organic solutes: (1) advection, (2) dispersion, (3) sorption and retardation, and (4) chemical/biological transformation.[5] The following brief discussion of these principles is intended to introduce the transport concepts and is not intended to be an in-depth discussion of each subject. For a more thorough review of the transport principles, consult the noted references.

Advection. Advection, the process by which solutes are transported by the bulk motion of flowing groundwater, is the dominant factor in the migration of dissolved organics in most groundwater environments. For relatively coarse-grained aquifers, in the absence of significant recharge or removal, such transport is usually horizontal. Advective flux in any dimension is described by Darcy's law, corrected for porosity:

$$V = \frac{K}{N_e} H$$

where V = velocity (defined as distance/time)
         K = hydraulic conductivity
         N_e = effective porosity
         H = gradient of hydraulic head

The advective flux of a contaminant at concentration C is given by

$$O_a = VC$$

Natural groundwater velocities can typically range from 1 m/yr to 1000 m/yr.

Dispersion. Organic contaminants in solution will spread as they move with the groundwater. This movement, called dispersion, is a result of molecular diffusion and mechanical mixing. Kinetic activity of the dissolved organics causes molecular movement from a zone of high concentration to a zone of lower concentration. Groundwater flow through the aquifer media results in mechanical mixing.[6]

Dispersion is commonly expressed as a linear function of velocity in the direction of flow:

$$D_x = a_x V$$

where D_x = the dispersion coefficient in direction x
V = mean groundwater velocity
a_x = characteristic length of dispersivity in direction x

Dispersive flux of a contaminant is described by the product of a dispersion coefficient and the concentration gradient as

$$Qd = Dx_i^d \frac{C}{X_i^d}$$

Sorption/Retardation. Sorption can occur when a solute interacts with aquifer solids encountered along the flow path through adsorption, partitioning, ion exchange, and other processes.[7] The higher the fraction of the dissolved organic sorbed, the more retarded is its transport. The degree of interaction depends on the characteristics of the dissolved organic, aquifer media, and chemical characteristics of the groundwater.

Chemical Transformation or Biotransformation. Transformation of an organic substance into another compound is the result of complex chemical and biological reactions. Hydrolysis and oxidation are the principal chemical reactions, while microbial degradation accounts primarily for biological impacts. It is believed that the biological transformations occur much more rapidly than chemical transformations, although for certain low level organics a prolonged microbial acclimation period may be required.[8] In addition, certain high-molecular-weight or chlorinated organics may resist biological transformation or may degrade at very slow rates.

Phase-Separated Transport. The transport of immiscible organic liquids can be divided into two distinct classifications: less-dense-than-water (transport of liquids that float on water) and denser-than-water (transport of liquids that sink in water). Common less-dense-than-water organics include gasoline and other petroleum distillates, while denser-than-water organics consist principally of halogenated hydrocarbons. The migration of an immiscible organic in the subsurface is largely governed by its density and viscosity.

The density of certain denser-than-water organics may result in a dominant vertical component of transport that is marginally affected by horizontal groundwater flow. The viscosity and surface wetting properties of less-dense-than-water organics will regulate their movement through the groundwater's capillary zone and upon the water table. Figure 2 illustrates the influence of the capillary zone on hydrocarbon movement. The migration of the hydrocarbon in Figure 2 becomes less restricted as a function of the vertical distance from the water table.

Vapor Transport

Many of the liquids contained in underground storage systems have volatile components that when exposed to standard atmospheric pressures and temperatures release these components into the vapor phase. Some vapors from liquids stored in the subsurface, such as gasoline and hexane, can cause fire, explosion, and safety hazards.

For organic vapors present in the subsurface to be detected in the vadose zone, movement from the saturated zone through the capillary zone or directly from the capillary zone to the detection or monitoring system must occur. The rates of vertical vapor transport are

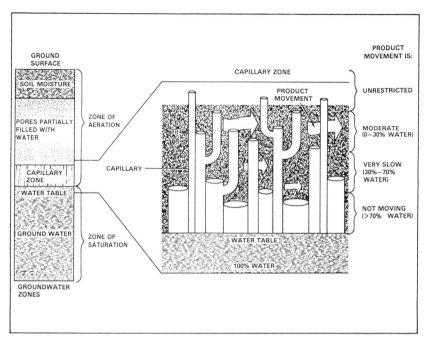

Figure 2. Capillary zone restricting hydrocarbon movement on the water table. (Source: API Publication 1628.)[3]

several orders of magnitude greater if the contaminant is present in and above the capillary zone.[9]

A fluctuating (recharging/discharging) water table above a contaminated aquifer may provide a more rapid mechanism by which organic vapors may move into the vadose zone. Figure 3 illustrates a simple case of a water table rising rapidly from Position 1 to Position 2. This rise pushed uncontaminated water in the capillary fringe upward into the vadose zone. When the water table falls, as shown in Position 3, contaminated water will be retained in the vadose zone and throughout the capillary zone. Hysteresis in the relationship between pressure head and water content enhances the retention of contaminated water in the vadose zone under these conditions of water table fluctuation. This enhancement occurs because, at a given tension, more water is retained in the pore spaces as the water table is lowered than enters the pores as the water table rises. This hysteresis in the pressure head/water content relationship is usually more pro-

RELATIVE GAS DIFFUSION COEFFICIENT

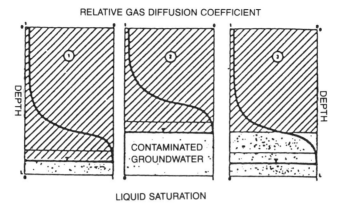

LIQUID SATURATION

Figure 3. Contamination of the vadose zone and capillary zone by a fluctuating water table. (Source: Lappala and Thompson, 1984.)[9]

nounced for coarse-grained soils near saturation than for fine-grained soils.[5]

Transport of organic vapors in the vadose zone is governed by the processes of diffusion and convection. Diffusion is the result of thermal motion of the vapor molecules subject to a concentration gradient. Convection results when a pressure gradient causes mass flow in the gaseous phase. Transport of the vapor phase in the vadose zone is a three-dimensional phenomenon. Contaminant flux caused by gaseous diffusion is described by Fick's first law, applied to a gas-filled pore space:

$$Q_g = D_g \frac{dC_a}{dz} \text{ with } D_g = \theta a(t) D_{ab}$$

where Q_g = mass flux per unit area per unit time
C = solute concentration
θa = the air-filled pore space
t = tortuosity
D_{ab} = the diffusion coefficient of gas a into gas b
dC_a/dz = concentration gradient

It has been determined that the opportunity for upward diffusion increases as drier soils are found closer to the land surface under conditions of insignificant recharge and redistribution of soil moisture.[9] In addition, vapor transport to low-pressure areas (basements, sewers) is a common problem during winter in cold regions. Frost at the ground surface prevents the upward diffusion of vapors, and hazardous situations can often result under these circumstances.

RELEASE DETECTION TECHNIQUES

Release detection techniques are commonly applied to assess the operational integrity of an underground storage system or to investigate a suspected release. These techniques have not been developed for long-term continuous monitoring application and are best suited for site-specific determination of the presence of a contaminant in the subsurface. The techniques that are described in this section can be readily implemented at most storage system sites. Depending on individual site characteristics, a combination of the techniques may be required to verify that the site possess no released substances or to determine the extent of the contamination and concentrations present.

Application of select release detection techniques in combination can provide the owner, operator, or manager of a storage system with a cost-effective release assessment program.

The discussion of release detection techniques has been divided into two subject areas: groundwater detection techniques and vadose zone detection techniques. The techniques applied in these two different subsurface regions use distinct liquid and vapor detection methodologies. The topics addressed under the heading of groundwater detection techniques are detection wells, soil sampling, dyes and tracers, and surface geophysics; topics addressed under vadose zone detection techniques are soil vapor sampling and remote infrared sensing.

Groundwater Detection Techniques

The groundwater beneath an underground storage system provides a ready medium to detect a release. In general, groundwater detection of leaks or losses is more readily determined in shallow- to moderate-depth groundwater systems (e.g., 0–50 ft). In shallow- to moderate-depth groundwater systems, leaked or lost organic substances from underground storage systems will, after overcoming the adsorbed capacity of the geologic matrix, either dissolve in groundwater, form an immiscible layer on the capillary zone (see Figure 2), or sink to the bottom of the aquifer. Where heavier-than-water organics of sufficient volume are lost, a phase-separated zone may accumulate below the top of the water surface on top of a lower aquilude or on the low-permeability geologic horizon.[10]

Detection Wells

Detection wells may be used to detect or define the movement and extent of contamination of a released substance in or on the groundwater. The use of detection wells is a well-established technology that provides an accurate visual confirmation of the subsurface conditions.

The steps in the design and construction of a detection well include:

1. Determine depth to groundwater.
2. Determine the depth of the well or monitoring point.
3. Select the appropriate drilling equipment for site-specific conditions.
4. Set up the drilling equipment.
5. Drill the bore hole.
6. Select compatible construction materials.
7. Install casings and liners.
8. Install well screens and fittings.
9. Gravel-pack the annular space between the screen and borehole.
10. Grout and seal the annular space.
11. Develop the well.
12. Establish either a manual or an automatic gauging system.

Depending on site-specific circumstances, all of the above steps may not be required for each well.

A variety of well drilling methods are available for the purposes of installing groundwater detection wells. It is important that selection of a drilling method be conducted to minimize the disturbance of subsurface materials and not introduce new contaminants into the groundwater system. Table 1 describes the various available drilling methods. The capabilities, costs, and advantages and disadvantages of the different drilling methods are detailed in Table 2.

Detection wells should be constructed from materials that are durable enough to resist chemical and physical degradation and do not interfere with the quality of the groundwater sample. A number of different well construction materials have been used including

Table 1. Well Drilling Methods[a]

Equipment Type	Procedure Description
1. Cable Tool	The cable tool drills by lifting and dropping a string of tools suspended on a cable. A bit at the bottom of the tool string strikes the bottom of the hole, crushing and breaking the formation material. Cuttings are removed by bailing or recirculation of a slurry. Casing is usually driven concurrent with drilling operations.[b]
2. Hydraulic Rotary Drilling	Hydraulic rotary drilling consists of cutting a borehole by means of rotating bit and removing the cuttings by a continuous circulation of a drilling fluid as the bit penetrates the formation materials.[c] Drilling fluid is usually a water-based slurry containing some fine material (clay) in suspension.
3. Reverse Circulation Drilling	Same as (2) except the fluids are forced downhole outside of the casing and returns up the inside of the casing. For (2) the route is down inside the casing and returning up outside the casing.
4. Air Rotary Drilling	Same as (2) only forced air becomes the drilling fluid.
5. Air-Percussion Rotary Drilling	Rotary technique in which the main source of energy for fracturing rock is obtained from a percussion machine connected directly to the bit.
6. Hollow-Rod Drilling	Similar to (1) except with more rapid and shorter strokes. Ball and check valve on bit allows water and cuttings to enter and travel up the casing (hollow-rod).
7. Jet Drilling	Similar to (6) except water is pumped down inside casing (hollow-rod) and returns up outside the casing.
8. Driven Wells	Pipe with a driving plug on the end is driven into the ground by repeated blows from a driving weight.
9. Hollow Stem Auger Drilling	Wells can be drilled in unconsolidated material and sampled by coring, with no drilling fluid or air required. Limited to shallow (less than 100 ft) unconsolidated material.

[a]Source: Canter, L.W., and Knox, R.C., *Ground Water Pollution Control* (Chelsea, MI: Lewis Publishers, Inc., 1986), p. 18.
[b]Campbell, M.D., and Lehr, J.H., "Well Cost Analysis."[13]
[c]Johnson Division, UOP Inc., *Groundwater and Wells*, fourth edition.[12]

EXTERNAL TANK LEAK DETECTION

Table 2. Evaluation of Well Drilling Methods[a]

Drill Type	Normal Dia. Hole	Max. Depth	Average Time Per Hole	Normal Expense	Advantages	Disadvantages
1. Rotary	4–20 in.	Unlimited	Fast	Expensive	1. Good for deep holes 2. Can be used in soils and relatively soft rock 3. Wide availability 4. Controls caving	1. Need to use drilling fluid 2. Potential bore hole damage with drilling fluid 3. Requires drilling water supply
2. Stem Auger	4–8 in.	30–50 ft	Fast under suitable soil conditions	Inexpensive to moderate	1. Widely available 2. Very mobile 3. Can obtain dry soil sample while drilling	1. Difficult to set casing in unsuitable soils (caving) 2. Cannot penetrate large stones, boulders or bedrock 3. Normally cannot be used to install recovery wells
3. Hollow Stem Auger	4–8 in.	30–50 ft	Fast under suitable soil conditions	Inexpensive to moderate	1. Good for sandy soild 2. Can set casing thru hollow stem 3. Very mobile 4. Can obtain dry soil samples and split spoon samples 5. Controls caving	1. Casing diameter normally limited to 2–3 in. o.d. 2. Cannot penetrate large stones, boulders, or bedrock 3. Limited availability 4. Normally cannot be used for recovery wells
4. Kelley Auger	8–48 in.	90 ft	Fast	Moderate to expensive	1. Can install large dia. recovery wells 2. Drills holes with minimum soil wall disturbance or contamination	1. Large equipment 2. Seldom available in rural areas 3. May require casing while drilling
5. Bucket Auger	12–72 in.	90 ft	Fast	Moderate to expensive	1. Can obtain good soil samples 2. Can install large dia. recovery wells	1. Hard to control caving 2. At times must use drilling fluid 3. Normally very large operating area required

Table 2. Continued

Drill Type	Normal Dia. Hole	Max. Depth	Average Time Per Hole	Normal Expense	Advantages	Disadvantages
6. Cable Tools	4–16 in.	Unlimited	Slow	Inexpensive to expensive	1. Widely available 2. Can be used in soil or rock	1. Slower than other methods 2. Hole often crooked 3. May require casing while drilling
7. Air Hammer	4–12 in.	Unlimited	Fast	Expensive	1. Fast penetration in consolidated rock	1. Inefficient in unconsolidated soil 2. Very noisy 3. Control of dust/air release 4. Excessive water inflow will limit use
8. Casing Driving (well point)	2–24 in.	60 ft	Slow to moderate	Inexpensive	1. Very portable—consolidated rock 2. Readily available	1. Caving can be severe problem 2. Limited depth 3. Greater explosive hazard during excavating into hydrocarbons
9. Dug Wells	Unlimited	10–20 ft	Fast	Inexpensive	1. Ready available 2. Very large dia. hole easily obtained	1. Caving can be severe problem 2. Limited depth 3. Greater explosive hazard during excavating into hydrocarbons

[a]Source: API Publication 1628.[3]

Teflon®,* steel, PVC, polyethylene, epoxy biphenol, and polypropylene.[11] Selection of an appropriate well construction material is based upon the substances of concern stored underground. Detailed discussions on well design and construction procedures are available in the literature.[12,13] A typical detection well is illustrated in Figure 4.

Depending on the substance being investigated, there are various types of detection wells that may be used to detect and define groundwater contamination:[14]

- a well, screened or open, over a single vertical interval
- a well cluster, consisting of two or more wells screened to sample at various depths
- a single well with multiple sample points or screens or packed zones

The advantages and disadvantages of these configurations are summarized in Table 3.

A well screened over a single interval is illustrated in Figure 4. This is the most commonly used detection well; however, a well screened over a single interval may not provide satisfactory information on the vertical distribution of certain released contaminants and volumes of contaminant release.

Detection well clusters completed or screened at varying depths are commonly used to define the vertical distribution of a release. As shown in Figure 5, each cluster comprises several closely spaced detection wells screened at various depths. Groundwater samples from these wells are representative of the water quality at different depths. Table 4 lists some of the factors influencing the number of wells required to define a release properly.

The final type of detection well is a single well with multiple sampling points. This well is screened and sealed at multiple locations throughout the depth of the well, so that samples may be removed at different groundwater depths.

Placement of any release detection well is critical. A false sense of security can be established if by improper placement a release was not detected. Trained professionals familiar with the many factors in detection well design, construction, and installation should be consulted before proceeding with a detection program.

Detection well sampling can be conducted through the use of a

---

*Registered trademark of E. I. du Pont de Nemours and Company, Wilmington, DE.

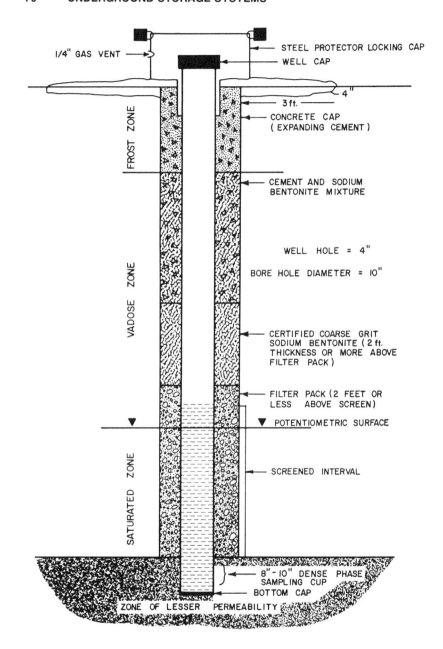

Figure 4. Typical detection well. (Source: EPA, "Draft RCRA Ground Water Monitoring Technical Enforcement Guidance Document," 1985.)[11]

Table 3. Types of Groundwater Detection Wells[a]

Type of Well	Advantages	Disadvantages
Well, Screened or Open, Over a Single Vertical Interval	Can provide composite groundwater samples if screen covers saturated thickness of groundwater table	No information given on the vertical spread of the contaminant
		Improper completion depth can cause error in determining the spread of contamination
		Screening over much of the aquifer thickness can contribute to vertical movement of contaminant
		The contaminant may become diluted in the composite sample, resulting in lower-than-actual concentrations
Well Clusters	Excellent vertical sampling made possible if sufficient number of wells are constructed	If only a few wells are installed, large vertical sections of the aquifer are not sampled. Artificial constraint on data by completion depths
	"Tried and true" methodology, accepted and used in most contamination studies where vertical sampling is required	Small diameter wells can be used only for monitoring. They cannot be used in abatement schemes
	Low cost if only a few wells per cluster are involved and if the drilling contractor has equipment suitable for installation of small-diameter wells (1–4 in. in diameter)	In small-diameter wells development and sample collection become tedious and difficult if water is below suction lift
Single Well-Multiple Sample Points (Nested Well)	Excellent information gained on vertical distribution of the contaminant	Relatively expensive
	If necessary, well diameter is large enough for use with pumping equipment	Proper well construction and sampling procedures are critical to successful application
	Sampling depths are limited only by the size and lift of the pump	It is possible to skip large sections of the groundwater table and thereby miss the contamination plume
	Rapid installation possible	

[a]Source: New York State Department of Environmental Conservation, "Technology for the Storage of Hazardous Liquids," 1983.[14]

number of different sampling devices ranging from very simple to extremely complex. To determine the presence or extent of a release accurately, proper sampling techniques must be observed. Selection of the appropriate sampling method requires considerable care and forethought to minimize the contribution of this process to the total

72 UNDERGROUND STORAGE SYSTEMS

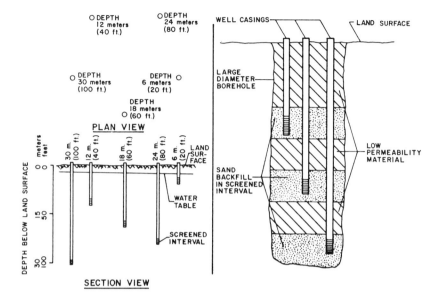

Figure 5. Typical detection well cluster. (Source: New York State Department of Environmental Conservation, "Technology for the Storage of Hazardous Liquids," 1983.)[14]

Table 4. Factors Affecting Number of Wells per Location (Clusters)[a]

One Well per Sampling Location	More than One Well per Sampling
• No "sinkers" or "floaters" (immiscible liquid phases)	• Presence of sinkers or floaters

• Thin flow zone (relates to 10-ft screen exception)	• Heterogeneous uppermost aquifer; complicated geology
	  – multiple, interconnected aquifers
	  – variable lithology
	  – perched water table
	  – discontinuous structures

• Homogenous uppermost aquifer; simple geology	• Discrete fracture zones

[a]Source: EPA, "Draft RCRA Ground Water Monitoring Technical Enforcement Guidance Documents," 1985.

bias during sample analysis. Significant factors and constraints in the selection of the sampling device include:[15]

- depth to standing/static water in the sampling installation
- hydraulic conductivity of the formation
- volume of water to be flushed/purged from the installations
- method of purging water from installation
- accessibility of the sites to be sampled
- parameters of interest

Table 5 lists the various subsurface/groundwater sampling methods and describes the advantages and disadvantages of each. The suitability of several sample collection techniques for specific parameters of interest is given in Table 6.

Soil Sampling

Soil core sampling can be an effective technique for determining the presence of an unconfined formerly stored substance in the subsurface. This detection technique is commonly employed in conjunction with the installation of a leak detection well. During installation of a soil boring, soil samples are routinely collected in drilling. The sample of soil is normally visually classified and the results recorded to create a well log. Portions of the retained samples are transferred to airtight containers for further laboratory analysis. The samples can then be field or laboratory analyzed for the parameters of interest. (Note: in dealing with compounds that present characteristics of high volatility, a sample may require quick freezing on dry ice in order to produce representative results.)

It is important that the results of soil core sampling and analysis be interpreted in conjunction with detection well sampling and analysis results to obtain meaningful detection data.

Dyes and Tracers

Dyes and tracers have been used in limited circumstances to verify the source of an underground storage system release. The technique consists of introducing a strong dye or tracer into or in the vicinity of a storage system suspected of being the source of the contamination and monitoring the point at which the release was first discovered for the appearance of the dye or tracer. A variety of dyes and tracers are available, including organic and fluorescent dyes, metallic tracers,

Table 5. Advantages and Disadvantages of Groundwater Detection Methods[a]

Advantages	Disadvantages
DOWNHOLE COLLECTION DEVICES	
General	
• Greater potential to preserve sample integrity than many other methods because water is not driven by pressure differences	• Most devices are unsuitable for flushing, because they provide only discrete, and often very small, volumes of water. This problem can be avoided by using another method, which may be disruptive, to flush the installation, prior to using the downhole collection device for sampling
Bailers	
• Inexpensive to purchase or fabricate and economical to operate. This may permit the assignment of one colleciton device for each installation to be sampled, thereby circumventing problems of cross-contamination	• Usually very time-consuming when used for flushing installations, especially when the device has to be lowered to great depths. It can also be very physically demanding on the operator when the device is lowered and raised by hand
• Very simple to operate, and require no special skill	• Can cause chemical alterations due to degassing, volatilization, or atmospheric invasion when transferring the sample to the storage container
• Can be made of inert materials	
• Very portable, and require no power source	
Mechanical Depth-Specific Samplers	
• Inexpensive to construct	• Some of the materials used can cause contamination (e.g., rubber stoppers)
• Very portable, and require no power source	• Activating mechanisms can be prone to malfunctions
• Stratified sampler is well suited for sampling distinct layers of immiscible fluids	• May be difficult to operate at great depths
• Can be made of inert materials	• Can cause chemical alteration when transferring sample to storage container
• Stratified sampler is easily cleaned	• Difficult to transfer sampler to storage container
	• Kemmerer sampler is difficult to clean thoroughly

Table 5. Continued

Advantages	Disadvantages
Pneumatic Depth-Specific Samplers	
• Can be made of inert materials	• Types that are commercially available are moderately expensive
• Easily portable, and require only a small power sources (e.g. hand pump)	• Westbay sampler is compatible only with the Westbay casing system
• Solinst sampler and syringe sampler can be flushed downhole with the water to be sampled	• Solinst and Westbay samplers are difficult to clean
• Syringe of the syringe sampler can be used as a short-term storage container	• Materials used in disposable syringes of syringe samplers can contaminate the water
• Syringe sampler is very inexpensive	• Water sample comes in contact with pressurizing gas in Solinst and Westbay samplers (but not in syringe samplers)
SUCTION-LIFT METHODS	
• Simple, convenient to operate, and easily portable	• Limited to situations where the water level is less than 708 m (23–26 ft) below ground surface
• Inexpensive to purchase and operate	• Can cause sample bias as a result of degassing and atmospheric contamination, especially if the sample is taken from an inline vacuum flask
• Easily cleaned	
• Components can be of inert materials	
• Depending on the pumping mechanisms, these methods can be very efficient for removing standing water from the sample installations	• Can cause contamination if water is allowed to touch pump components
• Provide a continuous and variable flow rate	
POSITIVE-DISPLACEMENT METHODS	
General	
• Reduced possibility of degassing and volatilization because the sample is delivered to ground surface situations (the pressure at ground surface may be substantially less than the natural water pressure in the formation and thus the gassing problem can not be entirely ignored)	• Cost of the commercially available pumps is substantial (roughly $2,000 to $5,000). It would therefore not be feasible to dedicate a sampling pump to each sampling point
• Sample does not contact the atmosphere	• Can be difficult to clean between sampling seasons

Table 5. Continued

Advantages	Disadvantages
• Sampling pumps for use in monitoring wells as small as 3.8–5 cm (1.5–2 in.) are commercially available	• Cleaning of cables and/or delivery tubing is required between sampling points
• Most of the commercially available devices have a sufficient flow rate for flushing sampling installations	• Commercially available devices are too large for very small-diameter installation such as the bundle piezometers

Submersible Centrifugal Pumps

Advantages	Disadvantages
• Can pump at large and variable flow rates	• Subject to excessive wear in abrasive or corrosive waters
• Johnson-Keck pumps can fit down wells as small as 5 cm (2 in.)	• Conventional submersible pumps cannot be used in installations of diameter less than about 12 cm (4 in.)
• Johnson-Keck pump is easily portable	
• Conventional pumps are usually much cheaper than the Johnson-Keck pump	• Potential for contaminating water because of contact with metals and lubricants is larger in conventional pumps
• Johnson-Keck pump offers little potential for sample contamination because it is made mostly of stainless steel and Teflon	• Johnson-Keck pump has intermittent flow (15 min on, 15 min off)

Submersible Piston Pump

Advantages	Disadvantages
• Gas-drive piston pumps have small power requirements	• Rod pumps require large power source, and are permanently mounted
• Gas-drive piston pump of Gilham and Johnson (1981) is inexpensive and can be assigned permanently to sampling point, thereby eliminating problems of cross-contamination	• Difficult to clean
	• When used as part of an installation, the gas drive pump of Gilham and Johnson (1981) cannot be retrieved for servicing or repair
• Double-acting pumps have continuous, adjustable flow rates	• Single-acting pumps have intermittent flow
• Can be built of inert materials (most commercially available pumps are not, however)	

Gas-Squeeze Pump

Advantages	Disadvantages
• Can be built of inert materials	• Intermittent but adjustable flow
• Commercially available pumps can be fit in installations as small as 5 cm (2 in.)	• Require large but portable power source
• Can easily be taken apart for cleaning, but can be inconvenient to clean between sampling sessions	

EXTERNAL TANK LEAK DETECTION

Table 5. Continued

Advantages	Disadvantages
• Easily portable	
• Good potential to preserve sample integrity because the driving gas does not come in contact with the water sample	

GAS-LIFT METHODS

Advantages	Disadvantages
• Simple to construct or are available commercially at relatively low cost	• Can be used efficiently when roughly 1/3 of the underground portion of the device is submerged
• Can be used in very narrow installations	• Contamination of the sample with the driving gas, atmospheric contamination, and degassing are all unavoidable
• Can be easily portable	• Need large power source (gas)
• Easily cleaned	

GAS-DRIVE METHODS

Advantages	Disadvantages
• Can offer good potential for preserving sample integrity, because very little of the driving gas comes in contact with the water sample, and because the sample is driven by a gradient of positive pressure	• Not very efficient for flushing installations larger than about 2.5 cm (1 in.)
	• Can be difficult to clean between sampling sessions
• Can be incorporated as part of the sampling installation, thereby removing the possibility of cross-contamination	• The driving gas comes in contact with the water, and therefore the beginning and the end of the slug of water obtained at the surface can be contaminated
• The triple-tube sampler is well suited for installations of very narrow diameter, e.g. 0.95 cm (3/8 in.), where the only other possible sampling method is narrow-tube bailers, or suction lift (when applicable)	• When used as part of a permanent sampling installation, they cannot be retrieved for repair or servicing
• Inert materials can be used	• Pump intermittently, and at variable flow rate

JET-PUMPS

Advantages	Disadvantages
• Can be used at great depths	• Use circulating water which mixes with the pumped water. A large amount of water needs to be pumped before the circulating water has a composition that is close to the water in the installation
• Useful for flushing monitoring installations	

Table 5. Continued

Advantages	Disadvantages
	• The water entering the venturi assembly is subjected to a pressure drop (which may be large), and can therefore undergo degassing and/or volatilization
	• The circulating pump at the surface can contaminate the pumped water because of its materials and lubricants
DESTRUCTIVE SAMPLING METHODS	
• Can provide very useful information in reconnaissance surveys and in other specific field situations	• Because no permanent installation is left in the ground, these methods cannot be used for monitoring long-term trends in water quality. In most cases, however, they do not interfere with the implantation of permanent installations
• Most of the techniques are used during the drilling operation and will not interfere with the implantation of a permanent installation	
• Coring-extraction methods are the only convenient means of obtaining several parameters related to both the liquid and solid phases (e.g. exchangeable cations, total microbial population, samples of the formation, etc.) and also for certain situations they may be the least bias-inducing method (e.g., in very fine-grained formations)	• Can result in large drilling costs
	• Water contained in cores can be contaminated with drilling fluids and can undergo degassing and volatilization at the ground surface
• Temporary installations can, in some situations, be the most cost-effective way of obtaining preliminary and/or reconnaissance data	

[a]Source: API Publication 4367.[15]

and radioactive tracers.[14] (Note: the use of colored dyes in geologic systems of high silt, clay, and organic content often proves of limited value, as the dye components are adsorbed onto the formation.)

Rhodamine B is a fluorescent dye, generally recommended for velocity and dispersion measurements. Fluorescein and rhodamine dyes have been used frequently in a variety of groundwater investigations. Freon (Cl_3CF) has been shown to yield positive results when used to determine groundwater velocities.[16] In determining rates of groundwater flow, it must be noted that normal rates are measured in units

EXTERNAL TANK LEAK DETECTION   79

Table 6. Suitability of Well Sampling Methods[a,b]

Parameters	Bailers	Downhole Collection — Depth-Specific		Suction-Lift Suction on:		Positive-Displacement								Destructive-Sampling		
						Centrifugal Pumps		Piston Pumps							Temporary Installations	
		Mechanical	Pneumatic	liquid phase	gas phase	Conventional	Johnson-Keck	Rod	Gas-Driven	Squeeze Pumps	Gas-Lift	Gas-Drive Pumps	Jet-Pumps	Coring-Extraction	Screened Augers	Multiple Completion Wells
INORGANIC																
Contamination Parameters																
Electrical condition	S	S	S	S	S	S	S	S	S	S	(U)	S	(U)	S	S	(L)
pH	(L)	(L)	S	S	(U)	S	S	(L)	S	S	(U)	(L)	(L)	(L)	S	(L)
Redox condition	(L)	(L)	S	(L)	(U)	(L)	S	L	S	S	(U)	(L)	L	(U)	(L)	(U)
Nontoxic Constituents																
Chloride	S	S	S	S	S	S	S	S	S	S	S	S	(L)	S	S	S
Sulfate	(L)	(L)	S	(L)	(L)	S	S	(L)	S	S	(U)	(L)	(L)	(L)	(L)	(L)
Sodium	S	S	S	S	S	S	S	S	S	S	S	S	(L)	S	S	S
Ammonium	(L)	(L)	S	S	S	S	S	S	S	S	(U)	S	L	(L)	(L)	(L)
Calcium, magnesium	S	S	S	S	(U)	S	S	S	S	S	(U)	S	(L)	(L)	(L)	(L)
Iron, manganese	(L)	(L)	S	(L)	(U)	S	S	(L)	S	S	(U)	(L)	(L)	(L)	(L)	(L)
Toxic Constituents																
Nitrate	S	S	S	S	S	S	S	S	S	S	(L)	S	(L)	(L)	S	(L)
Fluoride	S	S	S	S	S	S	S	S	S	S	(L)	S	(L)	(L)	S	(L)
Arsenic																
Selenium																
Barium																
Cadmium	(L)	S	S	(L)	(U)	S	S	(L)	S	S	U	(L)	(U)	(L)	(L)	(L)
Chromium																
Lead																
Silver																
Mercury																
RADIOACTIVE																
Radium	S	S	S	S	(L)	S	S	S	S	S	U	S	(L)	(L)	(L)	(L)
Gross alpha and beta	(L)	S	S	(L)	U	S	S	(L)	S	S	U	(L)	(L)	(U)	(L)	(L)

Table 6. Continued

	Downhole Collection					Positive-Displacement								Destructive-Sampling		
		Depth-Specific		Suction-Lift		Centrifugal Pumps		Piston Pumps						Coring-	Temporary Installations	Multiple
				Suction on:												Completion
Parameters	Bailers	Mechanical	Pneumatic	liquid phase	gas phase	Conventional	Johnson-Keck	Rod	Gas-Driven	Squeeze Pumps	Gas-Lift	Gas-Drive	Jet Pumps	Extraction	Screened Augers	Wells
BIOLOGICAL																
Coliform bacteria	S	(L)	S	S	(L)	L	(L)	U	(L)	S	(U)	(L)	(U)	S	(L)	(L)
ORGANIC																
Drinking-water stds. { Endrin, Lindane, Methoxychlor, Toxaphene, 2,4-D, 2,4,5-TP Silvex }	(L)	S	S	S	L	(L)	(L)	U	S	S	U	S	U	S	(L)	(L)
Quality Parameters																
Phenols																
Contamination Parameters { Total organic carbon, Total organic halogen }	(L)	(L)	S	(U)	U	(L)	(L)	U	(L)	S	U	(L)	(U)	(U)	(L)	(L)
Gasoline Components { Benzene, Toluene, Xylene, Methyl t-butyl ether }																

[a]Key: S: Suitable
L: Limited suitability; can be used for qualitative or approximate information
U: Unsuitable
( ): Procedural modifications can be used to improve suitability to next level

[b]Source: API Publication 4367.15

of a few tenths of a foot per day to a few ft/yr to maximum rates of a few tens of feet per day.

Techniques that use radioactive tracers as detection elements may also be used to pinpoint the source of an underground storage system release. Radioisotopes that have been applied to groundwater studies include H-3, I-131, Br-29, and Cr-EDTA.[7] However, a number of technical and public-relations concerns are associated with the use of radioactive tracers:

- A license or approval for use must be obtained from the U.S. Department of Labor or the Nuclear Regulatory Commission.
- Potential health and environmental hazards could result from the use of these materials.

The use of dyes and tracers for underground storage system release detection is often unsuccessful for several reasons:[17]

- Dye or tracer may contaminate the groundwater.
- If only vapor is found at the detection point, the dye or tracer will be useless.
- The dye or tracer may be adsorbed or oxidized by chemicals in the soil before it reaches the point of detection.
- As a result of slow groundwater flow, the testing may require considerable time.
- Dye or tracer may not be compatible with the stored substance of concern.

Surface Geophysics

Several surface geophysical techniques can be applied to detect and define the extent of a release from an underground storage system. The most common techniques include seismic refraction and electrical resistivity methods.[18]

The seismic refraction method is based on the fact that elastic waves travel through different earth materials at different velocities. The denser the material, the higher the wave velocity. When elastic waves cross a geologic boundary between two formations with different elastic properties, the velocity of wave propagation changes and the wave paths are refracted according to Snell's law. A source of seismic energy must be created, and the time between the energy shock and the arrival of the elastic wave is recorded.

The electrical resistivity of a geologic formation is defined as $p = R(A/L)$, where R is the resistance to electrical current for a unit block

Table 7. Typical Ranges of Geophysical Variables[a]

Method	Output Variables	Unit	Typical Range
Resistivity	Resistivity	ohm-m	$1-10^4$
	Sounding depth	m	0–100
Ground-penetrating radar	Depth of reflectors	m	<1–15
Electromagnetic induction	Conductivity	milliohm/m	0.10–1000
	Penetration depth	m	0.75–60
Metal detection	Penetration depth	m	0–3
Seismic refraction	Depth to refracting interface	m	1–30
Magnetometry	Magnetic field intensity gradient	gamma/m	0–300
	Magnetic field intensity anomaly	gamma	0–3000
	Ferromagnetic target depth (single drum)	m	0–5

[a]Source: Evans and Schweitzer, 1984.[18]

of cross-sectional area A and length L. The resistivity controls the gradients in electrical potential that will be set up in a formation under the influence of applied current. In an electrical resistivity survey, an electric current is passed into the ground through a pair of current electrodes and the potential drop is measured across a pair of potential electrodes. The spacing of the electrodes controls the depth of penetration. Sets of measurements are taken in the form of either lateral profiling or depth profiling.

The usefulness of geophysical techniques to detect and define the extent of a subsurface release depends upon several factors:[18]

- the depth of target contaminant plumes
- the contrast of contaminant plume conductivity with the background conductivity of nearby uncontaminated groundwater
- the intrinsic conductivity of the soil matrix
- the presence of overlying or underlying highly conductive zones, such as saline aquifers and clay lenses
- the presence of buried physical interferences (power lines, buried cables, sewers, and water mains)

Table 7 lists ranges of important variables for six geophysical methods. On the basis of this information it may be possible to estimate the likelihood of success when using these techniques to detect an underground storage system release.

Surface geophysical techniques have demonstrated promising results for inorganic contaminant detection. Currently, attention is being focused on applying these techniques to organic contaminant identification. Although surface geophysical techniques cannot replace test drilling, the use of the data generated by these techniques has the potential to reduce the number of detection wells required to fully define the extent of a subsurface release, and could result in considerable cost savings.

Vadose Zone Vapor Detection Techniques

As concern over releases from underground storage systems has escalated, there has been a renewed interest in an old technique of the oil and gas exploration industry, vadose zone (or soil) vapor analysis.[19] However, there are a number of vadose zone vapor detection techniques that concentrate primarily on identifying volatile organics and have been applied to underground storage system releases. Variations of these techniques focus on three aspects of vapor detection: (1) mode of probe design and construction, (2) vadose zone vapor sampling procedures, and (3) analytical techniques. This section reviews five vadose zone vapor techniques that have been applied to, or have the potential to be applied to, underground storage release detection: grab sampling of soil cores, surface flux chamber, downhole flux chamber, accumulator systems, and surface probe testing.[20]

As described earlier in this chapter, vapor transport in the vadose zone is governed by the process of molecular diffusion and convection. These two processes will be directly affected by the vadose zone geologic characteristics, groundwater table fluctuations, and meteorologic influences (e.g., precipitation, temperature, and atmospheric pressure).

Once vapors reach the ground surface, they are rapidly dissipated in the atmosphere. It is possible to perform ambient air monitoring and relate the observed data to soil gas emissions if meteorological conditions are known. However, comparison of direct and indirect (ambient) sampling techniques has demonstrated the superiority of direct vadose zone sampling.[21,22]

Grab Sampling of Soil Cores

Grab sampling of soil cores from the vadose zone involves collecting an undisturbed soil core — by use of an auger or by driving a hollow tube into the ground—and then sealing the core in a sample container with minimal headspace. Analysis of the headspace gas or extracted solids is performed at a later time. Table 8 reviews a number of different types of soil sampling equipment.

Gas chromatography is the principal analytical technique that has been applied to grab sampling of soil cores. Analysis for volatile organics can be broken down into three categories:[23] (1) organics in the pore space gas, (2) organics adsorbed to the soil particles, and (3) organics within the soil particles themselves.

Grab sampling of soil cores is relatively quick and simple. However, the method is better suited for measuring adsorbed organics and not free organics in the soil pore spaces. Because of the ease with which the technique can be applied, the approach is best suited as a crude release detection screening technique.

Surface Flux Chamber

The surface flux chamber technique is a direct detection approach that uses an enclosed system (flux chamber) to sample vapors from a defined surface area. Clean, dry sweep air is added to the chamber at a fixed, controlled rate. The volumetric flow rate of sweep air through the chamber is recorded and the concentration of the constituent of interest is measured at the exit of the chamber. Figure 6 is a diagram of a surface flux chamber.

A flux chamber should be instrumented with several devices to obtain usable data, including a source of high purity sweep air, air and soil temperature sensors, and analytical instrumentation. Portable and offsite gas chromatography is the primary analytical technique that has been used in conjunction with surface flux chambers.

The surface flux chamber technique is well suited to detect releases from underground storage systems if stainless steel evacuated gas canisters and a sufficiently sensitive gas chromatography system are employed. If portable gas chromatographs or other less sensitive analytical methods are used, the site soil should be unsaturated, without a heavy vegetative cover, and preferably consist of sand and silt layers.[20]

Downhole Flux Chamber

The downhole flux chamber operates on the same principles as the surface flux chamber; however, the system is designed to sample be-

Table 8. Criteria for Selecting Soil Sampling Equipment[a]

Type of Sampler	Obtains Core Sample		Most Suitable Core Types			Operation in Stony Soils		Most Suitable Soil Moisture Conditions			Access to Sampling Sites During Poor Soil Conditions		Relative Sample Size		Labor Requirements	
	Yes	No	Clay	Sand	Inter	Fav	Unfav	Wet	Dry	Inter	Yes	No	Sm	Lg	Sngl	2/More
A. Hand Auger																
1. Screw-type augers		X			X		X	X			X		X		X	
2. Barrel augers																
a. Posthole auger	X		X			X		X			X			X	X	
b. Dutch auger	X		X			X		X								
c. Regular barrel auger	X		X			X				X	X		X		X	
d. Sand augers	X			X		X				X	X			X	X	
e. Mud augers	X		X			X		X			X			X	X	
3. Tube-type samplers																
a. Soil probes																
(1) Wet tips		X			X		X	X			X		X		X	
(2) Dry tips		X			X		X	X			X		X		X	
b. Veihmeyer tubes		X			X					X	X		X		X	
c. Thin-walled tube samplers	X		X				X			X	X			X		X
d. Peat-samplers	X		X				X	X			X			X		X
B. Power Auger																
1. Hand-held screw type power auger		X			X	X				X	X			X		X
2. Truck-mounted auger																
a. Screw type		X			X	X		X				X		X		X
b. Drive sampler	X		X			X				X		X		X		X
3. Tripod-mounted drive sampler	X		X							X		X		X		X

[a]Source: API Publication 4394.[20]

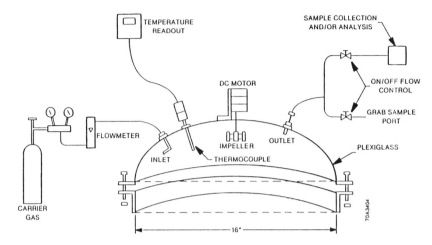

Figure 6. Surface flux chamber and peripheral equipment. (Source: API Publication 4394.)[20]

low the ground surface, and uses a smaller chamber that fits into a borehole. Sampling is usually conducted on freshly exposed soil with a hollow-stem auger as the borehole is augered. Figure 7 is a schematic diagram of a downhole flux chamber.

The analytical techniques applied to downhole flux chamber sample analysis are identical to those used with surface flux chambers. The downhole flux chamber technique has the advantage over surface flux chambers of being able to sample different levels of the vadose zone directly, and is suited for application to release detection from underground storage systems.

Accumulator Systems

The accumulator system technique involves static or dynamic collection and concentration of vadose zone vapors. The sampling time can vary with the type and amount of trapped vapor required for analysis. The technique provides an integrated sample that averages out any short-term fluctuations in vapor concentration.

A wide variety of sample gas extraction and accumulation procedures exist. In general, the procedures involve addition of an accumulator device to previously discussed sampling techniques. Figure 8 illustrates a Curie-point wire accumulator device.

Accumulator devices are inexpensive to operate and are relatively

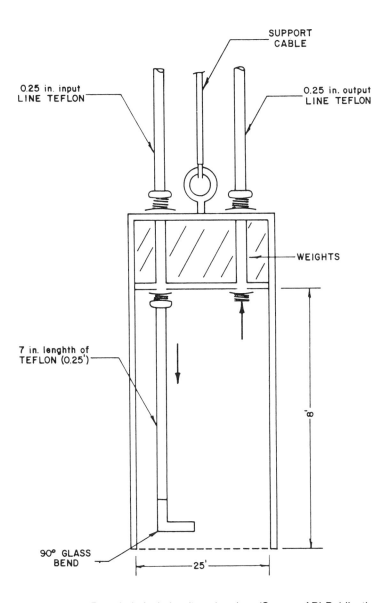

Figure 7. Downhole isolation flux chamber. (Source: API Publication 4394.)[20]

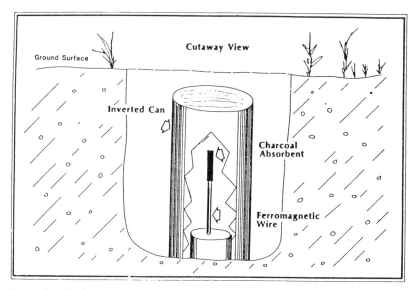

Figure 8. Curie-point accumulator device. (Source: API Publication 4394.)[20]

unaffected by weather and site conditions. However, they do require lengthy sample times and have unknown collection efficiencies (i.e., sample reaches accumulator by diffusion and the volume of gas sampled is unknown). Accumulator devices are suitable for detection screening of an underground storage release.

Ground Probe Testing

The ground probe technique involves the placement of a tube into the ground to the desired sampling depth. Vadose zone vapors enter the tube through openings near the tube's leading edge. The upper end of the tube contains an opening to permit sample gas to be extracted, usually via a gastight syringe. (See Figure 9.) Ground probes used to sample vadose zone vapors were reported as early as 1853.[24] The most common ground probe installations are conducted by passively placing the probe into a corehole and then backfilling, or by attaching a drive point to the end of the probe and driving the probe to the desired depth.

All studies using the ground probe technique since 1960 have reported using gas chromatography as the analytical technique; the lone exception is the use of detector tubes. These tubes are packed with a

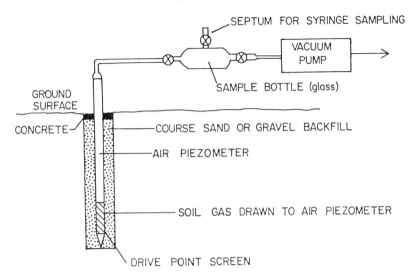

Figure 9. Driven probe soil gas sampling system. (Source: Thompson, 1985.)[25]

reagent that changes colors to indicate the concentration of the specific compound being tested. (See the section of this chapter on vapor wells for a more detailed discussion.)

Drivable ground probes are being used with increasing frequency to detect releases from underground storage systems. A dual-tracer release detection method designed to differentiate between liquid and vapor losses from underground storage systems has been developed.[25] This technique may have significant application to existing underground storage systems, a class of storage systems from which it is difficult to obtain definitive testing results because of the frequent presence of elevated background concentrations of the stored substance that can easily mask an active release. Figure 10 illustrates the application of dual-tracer ground probe detection techniques to underground storage systems.

Ground probe techniques have been successfully demonstrated at a variety of sample sites. It is a suitable vadose zone underground storage release detection technique, except where soils have high moisture content and low permeability, or where there are near-surface rock strata.

Table 9 lists the differences among the preceding five techniques for vapor sampling in the vadose zone.

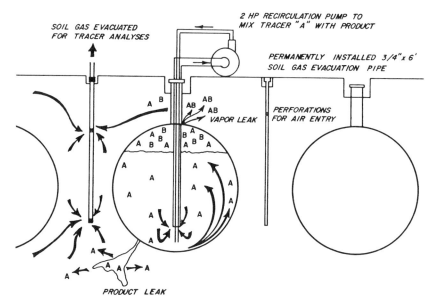

Figure 10. Dual-tracer leak detection method. (Source: Thompson, 1985.)[25]

Physical Inspection

Visual Inspection

Unless an underground storage system is maintained within an accessible vault, external visual inspections of the tank shell are not possible. However, if the tank is equipped with a manhole, internal inspections may be performed. Safety procedures must be strictly adhered to during any internal inspection of an underground tank. The tank must be emptied of liquid and any accumulated solids, properly freed of organic chemicals and vapors, and washed and cleaned before anyone enters the container for inspection. It is a sound safety precaution to wear a breathing apparatus and fire-resistant clothing when inspecting a tank that contained flammable materials.

Integrity Testing (Tank Shell)

Procedures to test the integrity of the tank shell are available, and include ultrasonic testing and radiation-type testing. Ultrasonic testing is used to measure the thickness of the tank shell and to determine

EXTERNAL TANK LEAK DETECTION 91

Table 9. Comparison of Vadose Zone Vapor Sampling Techniques[a]

Technique	Disturbance of Soil Gas Equilibrium		Suitable for Rocky Soil		Sampling Time		On-Site Analysis Practical		Sampling At Greater Depths If Necessary		Required Sampling Equipment		Required Analytical Equipment		Estimated Sampling Manpower	
	Small	Large	Yes	No	Hour(s)	Days	Yes	No	Yes	No	Simple	Complex	Variable	Complex	1	2
1) Grab Sampling of Soil Cores																
–Auger		X		X	X		X		X		X		X		X	
–Driven Sleeve		X		X	X		X			X	X			X	X	
2) Surface Flux Chamber	X		X		X		X			X		X	X			X
3) Subsurface Flux Chamber																
–Auger/Enclosure		X		X	X		X		X			X		X		X
–Groundprobe type	X		X		X		X		X			X		X		X
4) Accumulator Device																
–Curie-Point Wire	X		X			X		X		X	X			X	X	
–Absorbant/Pump		X	X		X		X		—	—		X		X	X	
5) Groundprobe																
–Passively Emplaced	X			X	X		X			X		X	X			X
–Driven	X		X		X		X		X		X			X	X	
–Driven, small volume	X		X		X		X		X		X			X	X	

[a]Source: API Publication 4394.[20]

the location, size, and nature of defects. Two types of instruments, pulse and resonance, are used.

The pulse type instrument transforms electric pulses into pulses of ultrasonic waves. The waves travel through the metal until a reflecting surface is reached. The waves then are reflected back, converted to electrical pulses, and displayed on the time baseline of an oscilloscope. The instrument is calibrated with a material of known thickness so that during measurement the time interval between the pulses corresponds to a certain shell thickness.

Two types of technologies are applied to the resonance type of measuring devices. The first technology uses an electronic oscillator which transmits electric energy of constant ultrasonic frequency to a crystal which then converts this energy into mechanical pressure waves that travel through the material being measured in the direction of its thickness. The pressure waves travel at a constant velocity and are reflected at the opposite surface back to the crystal. The time required for a wave to return to the crystal is a function of the distance traveled and can be converted to a measured thickness.

The second technology uses a crystal applied to the surface of the tank wall to be measured and an electronic circuit that causes the wall to vibrate over a range of frequencies. When the vibrating frequency of the crystal matches the natural frequency of the vibration of the material being measured, a signal is transmitted and interpreted electronically as an indicated thickness. This indication is conveyed to an oscilloscope tube and emerges as a series of vertical lines across the face of the tube, indicating the thickness of the material being tested.

Radiation-type integrity testing techniques principally use X-rays and gamma rays. An electromagnetic ray source is placed on one side of the tank shell and a photographic film is placed on the other side. When the rays pass through the object, the absorption of rays by imperfections is less than that of the solid material. Following development of the film, the imperfections will appear as darkened areas, approximately the size and shape of the flow.[14] Obviously, this technique is of highly limited use — it is applicable only to tanks that can be both internally and externally accessed.

The effectiveness and application of physical inspection to underground tanks is limited. Most underground storage tanks in the United States do not have manways and have little or no access to the external portion of the tank shell. In addition, the piping system generally cannot be physically inspected, and the frequency of inspection is limited by the time and cost of emptying and cleaning the

Table 10. Desired Properties of an External Release Monitoring System

- cost effective
- sensitive to small releases
- continuous or frequent operation
- reversible
- unaffected by normal background levels
- screen false alarms
- simple, safe, and reliable operation
- minimum maintenance requirements
- multiple monitoring locations
- applicable to both new and existing storage systems

container. However, in Europe (particularly in West Germany), periodic physical inspection of underground storage tanks by a governmental official is a standard component of all storage system management programs.[26]

RELEASE MONITORING SYSTEMS

Incorporation of some form of release monitoring is a critical element of any regulatory compliance or storage system management program. Monitoring for releases outside of the storage system is one method that can be applied continuously. The effectiveness and degree of confidence that an owner, operator, or manager should place in an external monitoring system depends on a number of significant factors. These factors include depth to water table, geologic characteristics, type of substance being stored, the sophistication of the monitoring technology being applied, and the location of the monitoring system relative to the storage system.

A number of the external monitoring technologies have had extensive field testing and have demonstrated successful release monitoring capabilities. Nevertheless, when there is more than one storage system at a facility, it is difficult (depending upon sampling locations) to determine specifically which system has failed. In addition, interfering substances or surface spills can cause false alarms.

Selection of a release monitoring system that will effectively monitor the environment adjacent to a storage system requires application of specific criteria. The desired properties of a release monitoring system are detailed in Table 10. The system should be sensitive and yet distinguish a true release from normal background interference, should operate with a reasonable frequency, and should monitor the subsurface environment over a defined area. The system should also be capable of being installed in both new and existing facilities, as close to the storage system as is possible at reasonable cost. The release monitoring systems discussed below address observation wells, vapor wells, and U-tubes.

Observation Wells

Observation wells (or monitoring wells) are designed to monitor the groundwater in the immediate vicinity of an underground storage system. These wells are most effective when there is a shallow water table beneath the facility.

There are reports, however, in which observation wells have effectively monitored underground storage systems with deep water tables. In those situations the wells have typically been installed within a highly permeable zone (e.g. tank hole excavation) surrounded by highly impermeable consolidated soils.[27,28]

The most important factors contributing to the effectiveness of an observation well monitoring system are:[29]

- design and construction characteristics of the observation well(s) (e.g., diameter, fabrication material, depth, location and size of perforations in the screened intervals)
- location and number of the observation well(s) (e.g., proximity to the storage system and potential direction of a released substance)
- depth to the water table
- physical and chemical properties of the geologic materials between the storage system and the observation well
- physical and chemical properties of the stored substance
- type of monitoring equipment used

Design

The most critical design factor for an observation well monitoring system is the depth from ground surface to the water table. The depth to groundwater will influence the type of drilling required, the amount of materials, the placement of the wells, and the total cost for the complete installation. Additional factors that need to be considered include the groundwater gradient and fluctuations. The gradient will directly influence the rate of transport of a released substance, while the amount of water table fluctuation can limit the monitoring devices that can be used to instrument an observation well. Certain types of equipment can tolerate only a limited amount of water table elevation change to be effective.

The number of wells required to successfully monitor an underground storage facility will vary depending on site-specific conditions (e.g., size of the storage facility, number of underground storage tanks, amount of piping and hydrogeologic conditions). When more than one tank is installed, the American Petroleum Institute recommends two wells be installed at opposite corners within the tank hole excavation. Many regulatory requirements, however, stipulate that four wells or more be installed within a tank hole excavation, depending on the number of tanks at a facility.

The use of observation wells to monitor piping systems is an impractical application. Piping lengths often are long, covering an extensive geographic area. To monitor a lengthy piping run adequately would require numerous wells, with resulting inordinately high cost for design, construction, installation, and sampling. More cost-effective approaches to piping system monitoring are available.

Construction

Observation wells are typically constructed from at least 2 in. i.d. Schedule 40 PVC pipe; however, both the type and size of an observation well will depend on the substance to be monitored and the type of monitoring equipment selected. Slotted casing having 0.020 in. maximum slots should extend from 2 ft below grade to the bottom of the well when installed within a tank hole excavation; when installed outside of the excavation, it should extend a few feet into the water table, accounting for seasonal water level fluctuations. Solid casing should extend from ground surface down to the slotted portion of the observation well.

When secondary containment is used, only one observation well within the tank hole excavation is necessary, designed and constructed similar to a well for use with primary containment storage systems.

The proper design and construction of observation wells requires the integration of geologic, hydrogeologic, and chemical information and is a relatively specialized technology. There are numerous references that provide excellent guidelines, including Johnson UOP (1975),[12] Barcelona et al. (1983),[30] and Campbell and Lehr (1977).[13]

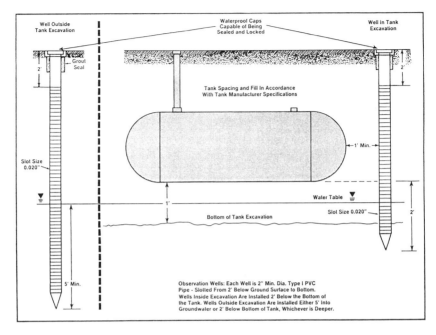

Figure 11. Typical observation well installations. (Source: API Draft, "Observation Wells as a Release Detection Technique," 1986.)[29]

Installation

An observation well is commonly installed within the tank excavation to a depth of 24 in. below the tank bottom and to the top of the concrete slab (if a slab is used to anchor the tanks in place).[28] Placement of an observation well outside of an excavation, a common practice in existing facilities, requires that the well extend a minimum of 5 ft below the water table or 2 ft below the tank bottom, whichever is greater. A typical installation at a retail petroleum facility is detailed in Figure 11.

Installation within a tank excavation is readily accomplished for a new storage system by excavating a limited volume of material from those locations in which the bottom of the well will be placed and then backfilling the entire excavation according to standard storage system installation procedures. Sloping the bottom of a tank excavation at a site with relatively consolidated natural soils and then placing a well at the lowest end can reduce the number of wells needed for effective monitoring of a new installation. The use of backfill material recommended by the storage system manufacturer will permit sufficient flow of groundwater into the observation well. Installation of an observation well to monitor an existing facility is more difficult than in a new facility and requires knowledge of the location of all buried structures, including tanks, piping, utilities, and sewer systems. Drilling for placement of an observation well can be accomplished through use of the techniques detailed previously in Table 1.

Sampling

A variety of approaches can be applied to sampling an observation well. A liquid sample can be removed from a well manually or automatically, and confirmation of the presence or absence of an unconfined substance can be performed visually or by sense of smell, laboratory analysis, or electronically onsite. Currently most of the observation well monitoring methodologies are specific to hydrocarbon; however, certain technologies are applicable to other organic materials, even though application has not been widespread. The following discussion will focus on the sampling of observation wells for hydrocarbon with specific reference to those technologies that can detect other organic materials.

Hydrocarbon-sensitive pastes. The use of a clean gauge stick or tape coated on one end with hydrocarbon-sensitive paste is an easy-to-use, inexpensive, and relatively sensitive manual method of monitoring an observation well. This technique is well suited for those operations that have available personnel onsite. A thin film of the paste is applied to one end of a gauge stick or tape, which is lowered into the observation well until the water table is contacted. A color change in the paste indicates that hydrocarbons are present within the observation well. The paste will indicate the presence of hydrocarbon at a minimum thickness of $1/64$ in. The frequency of sampling an observation well with this technique is determined by regulatory guidelines or by the facility owner/operator/manager. Sampling frequency will determine the rate of hydrocarbon-sensitive paste use. Cost estimates for paste use and purchase of a suitable tape are approximately $20.[29,31]

Bailers. The bailer (illustrated in Figure 12) is a portable tube with a ball valve at the bottom that is used to withdraw a sample of liquid from an observation well. If liquid phase hydrocarbons are present

98 UNDERGROUND STORAGE SYSTEMS

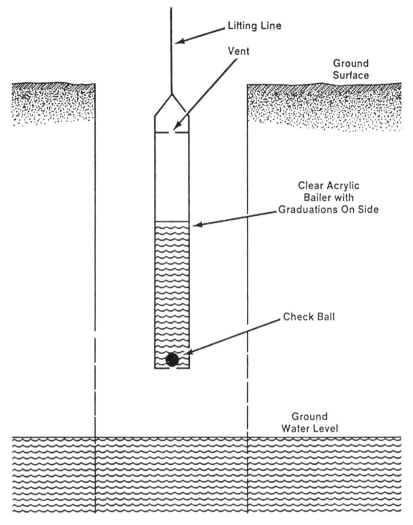

Figure 12. Schematic diagram of a bailer. (Source: API Draft, "Observation Wells as a Release Detection Technique," 1986.)[29]

within the observation well, the bailer will obtain a sample of the material that can be confirmed visually or by sense of smell. Concentrations of dissolved hydrocarbon as low as 0.1 ppm (threshold of odor) can be detected with a bailer.[32] In addition, laboratory analysis of the sample can be performed to monitor for the presence of other organic substances. (Note: only bailers of Teflon®, stainless steel, or other inert substances should be used for collecting samples intended for laboratory analysis.)

The use of bailers is a well-established sampling technology that is simple, inexpensive, sensitive, and reliable. The bailer is another technique that is well suited to those facilities with available personnel onsite. Training to obtain a representative sample from the observation well to monitor for the presence of the stored substance is minimal. Cleaning of the bailer should be conducted before sampling a different observation well to prevent cross-contamination and a resultant false reading. Bailers cost from $20 to $100 and are generally maintenance free.

Interface probes. An interface probe uses a conductivity sensor or a combination of refractive index/conductivity sensors to determine the presence of air, water, or hydrocarbon within an observation well. The system is calibrated so that the depth from ground surface to the liquid levels can be accurately recorded. An interface probe is a portable electronic system that is simple to operate and requires minimal maintenance. A minimum thickness of $1/32$ in. is detectable with an interface probe, and the system cost ranges from $1,400 to $2,000 each. This technology is applicable to substances denser than water.

Differential float device. A differential float monitoring system uses two different floats, one that floats on water and another that floats on any liquid lighter than water. The lighter-than-water-float is connected to a switch which activates an alarm. Figure 13 is a diagram of the differential float device. The device can detect a lighter-than-water material that has a minimum $1/16$-in. layer. The continuous monitoring device can instrument a 4-in. or a 12-in. observation well. The differential float device can be disrupted by debris or freezing conditions or if the float is dislodged. Each float assembly costs $500-$600 and the controller costs approximately $700. The controller can interface with upwards of six float assemblies.[29,31]

Thermal conductivity sensor. The thermal conductivity sensor is an automatic device designed to respond to the thermal conductivity differentials between air, water, and hydrocarbon. The system uses a floating probe and can detect a minimum of $1/8$ in. The devices have demonstrated good reliability during field evaluations provided that the probes are periodically cleaned. The system costs approximately

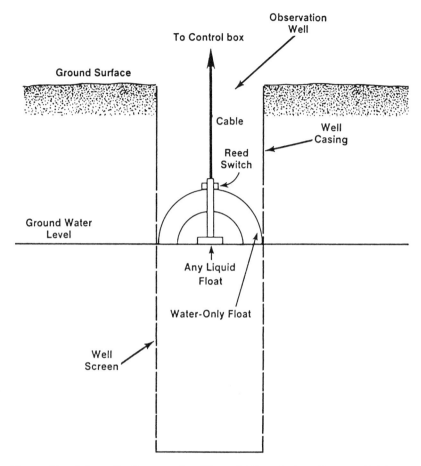

Figure 13. Schematic diagram of a differential float device. (Source: API Draft, "Observation Wells as a Release Detection Technique," 1986.)[29]

$700 per probe and approximately $1,500 for the required control unit.[29,31]

Electrical resistivity sensors. Electrical resistivity sensors use hydrocarbon-permeable coated wires that monitor the groundwater within an observation well. A change in the resistance of the wire will occur if the coating degrades as a result of contact with hydrocarbon. An electrical resistivity sensor is illustrated in Figure 14. The sensor is capable of detecting a released material at a minimum of $1/64$ in. The coating (a styrene-butadiene copolymer) is a relatively sensitive sub-

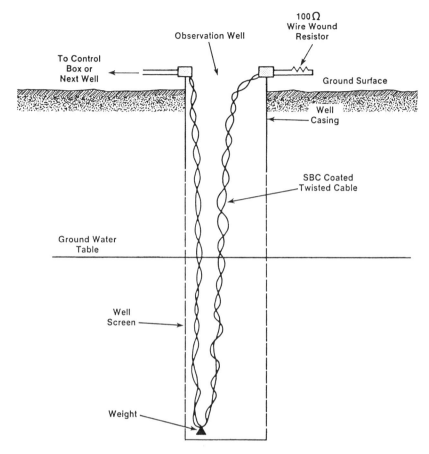

Figure 14. Schematic diagram of an electrical resistivity-sensing device. (Source: API Draft, "Observation Wells as a Release Detection Technique," 1986.)[29]

stance that is subject to degradation due to heat, abrasion, and ultraviolet rays. Prior contamination of the monitoring site or surface spills can also result in degradation of the coating. An electrical resistivity sensor will cost between $1,500 and $2,500 depending on the number of wells to be monitored.[29,31]

Hydrocarbon-soluble devices. A number of devices have been developed utilizing hydrocarbon-degradable materials (principally styrene-butadiene copolymer) to monitor for releases from an underground storage system. These devices are activated mechanically,

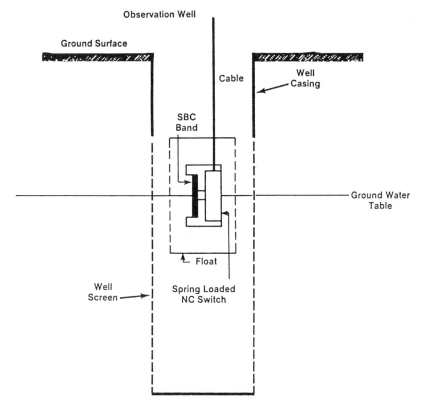

Figure 15. Schematic diagram of a hydrocarbon-soluble electrically activated device. (Source: API Draft, "Observation Wells as a Release Detection Technique, 1986.)[29]

electrically, or by pressure and often require the replacement of the detection mechanisms following activation. An electrically activated device is illustrated in Figure 15. The hydrocarbon-soluble devices have limitations similar to those outlined for electrical resistivity sensors (they are sensitive to heat, abrasion, ultraviolet rays, and prior site contamination). Costs for the devices are variable; the mechanical system is approximately $200 per well, the electrically activated system is approximately $8,000 for a four-well multifunctional system, and the pressure system is approximately $1,200 per well. Maintenance costs are not expected to exceed $50–$100 per year.[29,31]

Hydrocarbon-permeable devices. These are designed to detect hydrocarbons in an observation well through use of water-repellent but hydrocarbon-permeable materials. One application involves a constant-resistance tube sealed in a hydrocarbon-permeable membrane. The tube is mounted on a float and will change resistance when in contact with hydrocarbon. Another application uses a specially coated tube monitored by a series of sensors. These hydrocarbon-permeable devices must be cleaned periodically and replaced if activated by a release. Costs for these systems range from $2,000 to $5,000 per facility.[29,31]

Vapor Wells

Monitoring an underground storage system for a release by means of vapor detection is a relatively new monitoring technique. The methodology has demonstrated promise for application to newer facilities that have a water table deeper than 40–50 ft below the ground surface. For deep water table situations, groundwater monitoring is not practical for early release detection. Vadose zone monitoring, by means of vapor wells, is particularly applicable to storage systems containing highly volatile materials.

The long-term reliability of vapor monitoring for underground storage systems has yet to be fully determined. A greater database must be assembled documenting the application of vapor monitoring systems under varying operating conditions.

The effectiveness of a vapor well is influenced by a number of the same factors that affect observation wells, including:

- design and construction characteristics of the vapor well(s) (e.g., diameter, fabrication material, depth, location and size of vapor intakes)
- location and number of vapor well(s) (e.g., proximity to the storage system and potential direction of the release)
- physical and chemical properties of the geologic materials between the storage system and the vapor well(s)
- physical and chemical properties of the stored substances
- type of monitoring equipment used
- atmospheric conditions (barometric pressure, precipitation, wind velocity, temperature, etc.).
- depth and fluctuations of the water table

A significant deterrent to widespread application of vapor well monitoring for existing storage systems is the presence of background levels of the substance being monitored. Surface spills, prior releases, repairs, bacterial activity, or non-vapor-tight systems can affect the operation of a vapor well system and can result in frequent false alarms. However, many of the different vadose zone monitoring systems applicable to vapor wells have adjustable detection levels that can account for background concentrations of the substance of concern.

The design, construction, installation, and operation of a vapor well will also have a significant impact on the ability of the vapor well to detect a release. These subjects are discussed in the following sections.

Design

The most critical aspect in the design of a vapor well monitoring system is the site-specific geologic characteristics. The geologic conditions will determine the number of vapor wells, the depth of installation and installation technique, and type of material used. In addition, the geologic conditions will govern the vapor transport characteristics of an unconfined substance. Less permeable geologic conditions will require a greater number of vapor wells placed at greater depths. The size of the storage facility and the number of tanks and length of piping will also influence the number and placement of vapor wells.

Because application of vapor well monitoring to underground storage systems is so new, there are no readily available design, construction, or installation standards or recommended practices.

Construction/Installation

Vapor wells are typically constructed from 2-in.-diameter Schedule 40 PVC, or ³/₄-in. to 1¹/₄-in. carbon steel, or hollow, flexible tubing placed into an augered, backfilled hole.[24,33] The augered hole or well is typically capped with a tapered plug to assure that a seal at the ground surface is attained. The intake of the vapor well is usually located at the lower end of the well and can consist of vertical or horizontal slots or perforations. The depth of the well will vary from the bottom of the excavation to within several inches of the ground surface. Figure 16 is a schematic diagram of a vapor well.

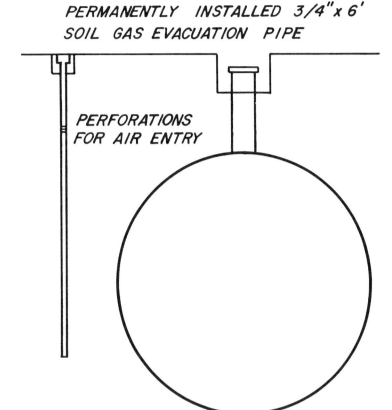

Figure 16. Vapor well monitoring system. (Source: Thompson, 1985.)[25]

Sampling

Vapor wells can be sampled by a number of different techniques, including detector tubes, combustible gas indicators, catalytic detectors, metal oxide semiconductors, photoionization detectors, and flame ionization detectors.[29] These sampling systems can be portable or dedicated.

Detector tubes. Detector tubes are small, portable glass tubes filled with reactive materials that change colors when the vapor of concern is aspirated through the tube. The tube is calibrated so that the greater the amount of reactive material that changes color, the higher

106 UNDERGROUND STORAGE SYSTEMS

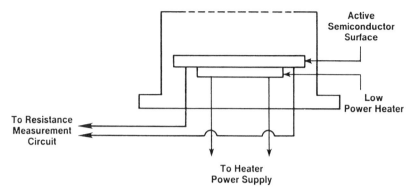

Figure 17. Schematic diagram of a metal oxide semiconductor. (Source: API Draft, "Observation Wells as a Release Detection Technique," 1986.)[29]

the concentration of vapor in the well. The sensitivity of the tubes will depend on the vapor being sampled. The tubes are relatively inexpensive (approximately $3 per tube), require little operator training to use, and are disposable after each use.

Combustible gas indicators (CGIs). CGIs are portable monitoring devices designed to measure combustible vapors and are in common use by the fire service. The device operates by aspirating a known volume of air into the device, where it comes in contact with a heated filament. The filament is linked to a resistance circuit and will indicate the presence of a vapor by a change in filament temperature that results in resistivity changes. A CGI is easy to operate, but does not have a wide sensitivity range. Maintenance for these devices is low, and each one costs approximately $300–$800.

Catalytic detectors. Catalytic detectors operate on the same principle as the CGI, but are more sensitive as a result of the use of two heated circuits and an amplifier circuit. Detection for combustible vapors ranges from 10 ppm to as high as 20,000 ppm. The catalytic detector is easy to use, portable, requires little maintenance, and can be purchased for $700–$1,500.

Metal oxide semiconductors. Metal oxide semiconductors (illustrated in Figure 17) monitor vapors by reacting to changes in internal electrical resistance. The devices are calibrated for the vapor of concern and can be portable or fixed. The dedicated system requires a built-in pump to draw air from the vapor well to the panel-mounted sensor. These systems are more complex than the previously mentioned systems; however, continuous multichannel monitoring can be achieved. Metal oxide semiconductors cost approximately $1,000 (for a single-channel system) to $4,000 (for a multichannel system).

Photoionization detector. The photoionization detector is a portable trace gas analyzer use to measure a number of organic compounds in multiple applications. These devices are highly sensitive; however, interference can be caused by excess water vapor. Photoionization devices are relatively sophisticated, require trained operation, and cost from $7,000 to $15,000.

Flame ionization detector. The flame ionization detector is a very sensitive instrument (ppm) which uses hydrogen gas to oxidize organic vapors. However, the instrument is relatively complex, requires specialized operation, and is not readily portable. An external hydrogen source is required to achieve effective operation. The flame ionization detector costs approximately $12,000 per unit and requires periodic maintenance.

U-Tubes

U-tubes have been used to monitor underground storage systems in only a few locations in the U.S. (particularly Long Island, New York). The system is limited to monitoring for releases from new underground storage tanks and is not applicable to the piping. The U-tube technology assumes that a loss of material from an underground storage tank will travel downward directly beneath the tank and enter the U-tube. The technique may be most appropriate for monitoring of storage systems in areas of low water tables.

There is limited operating experience with U-tubes, and no definitive data regarding the effectiveness of the technique. Factors that will directly affect the effectiveness of a U-tube include:

- proper installation
- depth of the water table
- amount of water table fluctuations

The U-tube must be positioned directly beneath the tank during installation, and the system is prone to flooding by entering rainwater or a high/rising water table.

108 UNDERGROUND STORAGE SYSTEMS

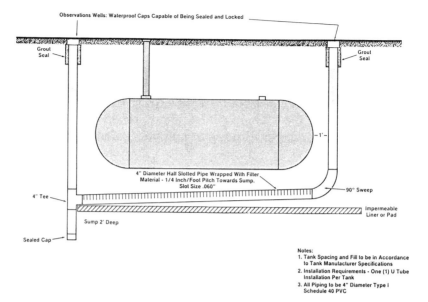

Figure 18. Schematic diagram of a U-tube installation. (Source: API Draft, "Observation Wells as a Release Detection Technique," 1986.)[29]

Design

A U-tube typically consists of a 4-in.-diameter Schedule 40 PVC pipe installed beneath the storage tank as shown in Figure 18. The horizontal segment of the pipe is slotted on the upper portion of the pipe with openings, typically 0.06 in. The pipe is wrapped with a mesh cloth to prevent backfill infiltration and sloped (approximately 1/4 in./ft) toward the sump. At the higher end of the pipe, there is a 90° sweep to a vertical pipe that is extended to grade. At the lower end of the horizontal pipe, there is a tee connection, with a vertical section extended to grade and 2 ft below the tee to act as a collection sump. All vertical pipe sections are unperforated, and the bottom of the sump is sealed. An impermeable barrier is placed beneath the U-tube to assist in the collection of any released substances. All openings to grade should have watertight, locked caps that are clearly identified as U-tube openings.

Construction/Installation

As with other forms of PVC casing, the U-tube can be assembled with slip joint fittings or threaded PVC. A glue, compatible with the liquid being stored, should be selected when slip fittings are used. U-tubes can be installed only at new installations before the tanks are placed in the excavation. The tubes can be installed beneath each tank in an excavation or centrally located. In either case, the excavation should be sloped toward the U-tube to permit collection of any unconfined material. A minimum of 12 in. of backfill should separate the U-tube from the bottom of the storage tank.

Sampling

U-tubes can be sampled using similar techniques or equipment to those previously described for application to observation well sampling. The tubes can be sampled manually with some form of portable sampling device, or automatically with a fixed system.

SECONDARY CONTAINMENT MONITORING

Under certain state and local regulatory programs (e.g., in California, Rhode Island, Maine, New Hampshire, and Austin, Texas) and in certain environmentally sensitive areas, new underground storage system installations should be secondarily contained. Secondary containment is simply another barrier that a release must pass through to reach the external environment. This additional barrier (secondary containment) typically consists of double-walled storage tanks or an impervious excavation liner. Other forms of secondary containment are concrete vaults, low-permeability clay liners, and soil cement liners.

Release detection or monitoring within a secondary containment system uses technologies similar to those applied to monitoring for a loss outside an underground storage system, except that losses can be detected more rapidly, since the sensing methodologies are employed in a confined space. Within an impervious excavation liner, concrete vault, low-permeability clay liner, or soil cement liner, any of the release detection and monitoring technologies described previously in this chapter as applicable to an observation well–based system is suitable for monitoring secondary containment. The choice of monitor-

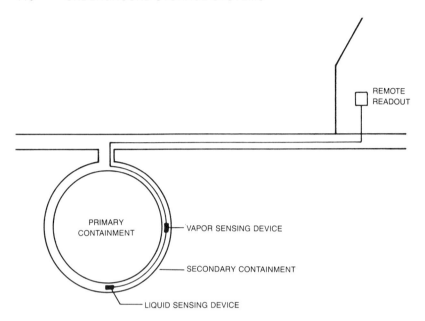

Figure 19. Typical Type 2 secondary containment sensor installation.

ing system depends on whether the secondary containment will be operated dry or with water in the liner/vault.

Release detection and monitoring of a double-walled tank can include alternative methodologies to those previously discussed, because the monitoring area for a double-walled tank — the space between the inner and outer tank shells — is sealed during fabrication. This sealed interstitial space permits employment of several active monitoring technologies, including a positively or negatively pressured interstitial space and a liquid-filled interstitial space.[34] Passive monitoring technologies include vapor and liquid sensing. Figure 19 illustrates the use of continuous electronic liquid and vapor monitoring devices in a double-walled tank. For illustrative purposes, both devices are shown within an interstitial space.

Positive or negative pressure of the interstitial space can be monitored visually with a pressure gauge or with continuous electronic devices and alarms. Typically, pressure of 2–3 psi within the interstitial space is maintained with an inert gas such as argon or with nitrogen. Several variables can affect the performance of a positive- or negative-pressure interstitial monitoring system: temperature fluctuations, atmospheric pressure changes, and tank loading pressures from tank filling and surface loading (e.g., heavy vehicles).

In a liquid-filled system, the interstitial space is filled with a brine—ethylene glycol, or water containing a corrosion inhibitor—and connected to an elevated reservoir/controller for monitoring releases. In this manner, a positive head pressure greater than any anticipated external water pressure is maintained on the liquid within the interstitial space.[35] If the integrity of either the inner or outer tank shell is breached, the liquid in the reservoir drops below a preset level and sets off an alarm. Depending on the liquid level in the tank, the tank contents and the monitoring liquid can mix.

As the use of secondary containment underground storage systems becomes more common, the effectiveness of these release detection and monitoring systems will be determined more reliably.

SUMMARY

External tank release detection and monitoring is relatively inexpensive and easily implemented when compared to other release detection and monitoring methodologies. The principal factors that affect external tank release detection and monitoring are:

- geologic and hydrogeologic conditions
- distance between the detection and/or monitoring system and the release source
- number and placement of detection and/or monitoring system
- physical and chemical properties of the stored substance
- type of detection and/or monitoring system used

To design, construct, install and operate an effective release detection or monitoring system requires a working knowledge of subsurface liquid and vapor transport. The transport of these substances is controlled by the type of substance released, the rate and volume of the release, and the surrounding hydrogeologic environment. The four primary physical principles that regulate liquid movement in the subsurface are (1) advection, (2) dispersion, (3) sorption and retardation, and (4) chemical/biological transformation. The processes of diffusion and convection govern subsurface vapor transport.

External tank release detection techniques are designed primarily for identifying the presence of an unconfined substance and are not designed for long-term monitoring applications. Groundwater release detection can be accomplished by the use of detection wells, soil sampling, dyes and tracers, and surface geophysics. Vadose zone detection techniques include soil vapor sampling and remote infrared sensing. Cost-effective application of the various detection techniques depends on site-specific conditions and the availability of equipment. Physical inspection of the accessible portions of a facility and, if possible, the interior of the tank is a further means of leak detection.

External tank release monitoring is a critical element of any regulatory compliance or storage system management program. A number of different techniques can be used to monitor new or existing underground storage systems, including observation wells, vapor wells, and U-tubes. Design, construction, installation, and sampling procedures will vary for each system. The effectiveness of a release monitoring system depends upon many factors:

- location of the monitoring system relative to the storage system
- geologic and hydrogeologic characteristics
- substance being stored
- sophistication of the monitoring/sampling technology being applied
- background subsurface concentrations of the stored substance

Installation of a release monitoring system within the tank excavation will improve the effectiveness of the system and reduce the impact of the many transport variables. As the choice of continuous and periodic monitoring systems continues to increase, evaluations defining the reliability, accuracy, and cost-effectiveness will have to be conducted. Initial efforts[36,37] have demonstrated that none of the monitoring technologies is free of problems or is applicable in all situations.

EPA Studies at the Environmental Monitoring Systems Laboratory in Las Vegas, Nevada, are focusing on developing release monitoring performance standards and evaluating various monitoring technologies.

REFERENCES

1. Schwendeman, T. G., "Transport and Recovery of Hydrocarbon in the Subsurface Environment." Philadelphia, PA: *Proceedings of the Third Annual Hazardous Materials Management Conference*, June 4–6, 1986.

2. American Petroleum Institute, "The Migration of Petroleum Products in Soil and Ground Water" (Publication 4149). Washington, DC: American Petroleum Institute, December 1972.
3. American Petroleum Institute, *Underground Spill Clean-up Manual* (Publication 1628). Washington, DC: American Petroleum Institute, June 1980.
4. Camp, Dresser and McKee, Inc., "Fate and Transport of Substances Leaking from Underground Storage Tanks: Interim Report" (Contract No. 68-01-6936). Washington, DC: Office of Underground Storage Tanks, U.S. Environmental Protection Agency, 1986.
5. Mackay, D. M.; Robert, P. V.; and Cherry, J. A., "Transport of Organic Contaminants in Groundwater," *Environ. Sci. Technol.* 19:5 (1985).
6. Anderson, M. P., *Movement of Contaminants in Groundwater: Groundwater Transport—Advection and Groundwater*. Washington, DC: National Academy Press (Studies in Geophysics: Groundwater Contamination), 1984.
7. Freeze, R. A., and Cherry, J. A., *Groundwater*. Englewood Cliffs, NJ: Prentice-Hall, Inc., 1979.
8. McCarthy, P. L.; Rittman, B. E.; and Bouwer, E. J., "Microbiological Processes Affecting Chemical Transformations in Groundwater." In Bitton, G., and Gerba, C. (eds.), *Groundwater Pollution Microbiology*. New York: John Wiley & Sons, Inc., 1984.
9. Lappala, E. G., and Thompson, G. M., "Detection of Groundwater Contamination by Shallow Soil Gas Sampling in the Vadose Zone: Theory and Applications." Washington, DC: National Conference on Management at Uncontrolled Hazardous Waste Sites, November 1984.
10. Hillel, D., *Soil and Water: Physical Principles and Processes*. New York: Academic Press, 1971.
11. Environmental Protection Agency, "Draft RCRA Ground Water Monitoring Technical Enforcement Guidance Document." Washington, DC: Office of Waste Programs Enforcement, U.S. Environmental Protection Agency, August 1985.
12. Johnson Division, UOP Inc., *Groundwater and Wells* (fourth edition). St. Paul, MN: UOP Inc., 1975.
13. Campbell, M. D., and Lehr, J. H., "Well Cost Analysis." In *Water Well Technology* (fourth edition). New York: McGraw-Hill Book Company, 1977.

14. New York State Department of Environmental Conservation, "Technology for the Storage of Hazardous Liquids: A State-of-the-Art Review." Albany, NY: Bureau of Water Resources, Division of Water, New York State Department of Environmental Conservation, January 1983.
15. American Petroleum Institute, "Groundwater Monitoring and Sample Bias" (Publication 4367). Washington, DC: American Petroleum Institute.
16. Thompson, G. M.; Hayes, J. M.; and Davis, S. N., "Fluorocarbon Tracers in Hydrology," *Geophysics Research Letters* 1 (1974).
17. National Fire Protection Association, "Underground Leakage of Flammable and Combustible Liquids" (NFPA 329). Quincy, MA: National Fire Protection Association, 1983.
18. Evans, R. B., and Schweitzer, G. E., "Assessing Hazardous Waste Problems," *Environ. Sci. Technol.* 18:11 (1984).
19. Horwitz, L., "Geochemical Extraction for Petroleum," *Science* 229, No. 4716:821-27 (August 30, 1985).
20. American Petroleum Institute, "Detection of Hydrocarbons in Groundwater by Analysis of Shallow Soil Gas/Vapor" (Publication 4394). Washington, DC: American Petroleum Institute, May 1985.
21. Eklund, B. M., and Schmidt, C. E., "Review of Soil Gas Sampling Techniques: Soil Gas Sampling Techniques of Chemicals for Exposure Assessment – Interim Report" (EPA-EMSL 68-02-3512, Work Assignment 32). Radian Corporation, 1983.
22. Balfour, W. D., and Schmidt, C. E., "Sampling Approaches for Measuring Emission Rates from Hazardous Waste Disposal Facilities." San Francisco, CA: Seventy-seventh Annual Meeting of the Air Pollution Control Association, June 24-29, 1984.
23. Hanish, R. C., and McDevitt, M. A., "Protocols for Sampling and Analysis of Surface Impoundments and Land Treatment/Disposal Sites for VOC's: Technical Note" (EPA-EMB 68-02-3850, Work Assignment 11). Washington, DC: U.S. Environmental Protection Agency, 1984.
24. Russel, E. J., and Appleyard, A., "The Atmosphere of the Soil: As Composition and the Causes of Variation," *J. Ag. Sci.* 7, Part 1 (1915).
25. Thompson, G. E., "Tracer Leak Detection Technology." Tucson, AZ: Tracer Research Corporation, 1985.

26. Personal communication with Marcel Moreau, geologist, Field Services Division, Bureau of Oil and Hazardous Materials Control, Department of Environmental Protection, Augusta, ME, July 1986.
27. Personal communication with Sully Curran, Staff Advisor— Environmental Legislation and Regulation, Marketing Department, Exxon Company USA, Houston, TX, June 1985.
28. American Petroleum Institute, "Recommended Practice for Underground Petroleum Product Storage Systems at Marketing and Distribution Facilities," second edition (Publication 1635). Washington, DC: American Petroleum Institute, December 1984.
29. American Petroleum Institute, "Observation Wells as a Release Detection Technique: Draft Research Report." Washington, DC: American Petroleum Institute, April 1986.
30. Barcelona, M. J.; Gibb, J. P.; and Miller, R. A., "A Guide to the Selection of Materials for Monitoring Well Construction and Groundwater Sampling" (SWS Contract Report 327). Champaign, IL: Illinois State Water Survey, Department of Energy and Natural Resources, 1983.
31. Scheinfeld, R. A.; Robertson, J. B.; and Schwendeman, T. G., "Underground Storage Tank Monitoring: Observation Well Based Systems," *Ground Water Monitoring Rev.* 6, No. 4 (Fall 1986).
32. American Petroleum Institute, "Review of Published Odor and Taste Threshold Values of Soluble Gasoline Hydrocarbons" (Publication 4419). Washington, DC: American Petroleum Institute, December 1985.
33. Spittler, T. M., and Clifford, W. S., "A New Method for Detection of Organic Vapors in the Vadose Zone." Lexington, MA: U.S. EPA Region 1 Laboratory, 1985.
34. Personal communication with Charles T. Erdman, quality engineer, Buffalo Tank Corporation, Baltimore, MD, November 1986.
35. Owens-Corning Fiberglas Corporation, "Double Wall Tank Installation Instructions." Toledo, OH: Owens-Corning Fiberglas Corporation, 1986.
36. Donaghey, L. F., "Groundwater Protection through Early Detection of Hydrocarbon Leaks." Los Angeles, CA: Proceedings, 1985 Oil Spill Conference, February 25–28, 1985.
37. Lu, J., and Barcikowski, W., "Cost Effectiveness of Leak Detection and Monitoring Technologies for Leaking Underground

Storage Tanks." Atlanta, GA: *Proceedings of the National Conference on Hazardous Waste and Hazardous Materials— Hazardous Materials Control Research Institute*, March 4-6, 1986.

CHAPTER 4

In-Tank Leak Detection Methodologies

H. Kendall Wilcox

CHAPTER CONTENTS

Basic Principles of Tank Testing 120
    Volumetric Testing 122
    Temperature Compensation Techniques 126
    Product Level Measurements 129
    Calibration of Level Sensor 134
    Determination of the Coefficient of Expansion 135
Variables in Volumetric Testing 136
    Temperature Effects 137
    Vapor Pockets 140
    Water Table Effects 143
    Tank Distortions 144
    Vibration 146
    Evaporation and Condensation 147
    Head Pressure Effects 147
Nonvolumetric Testing 149
    Helium Leak Detection 150
    Tracer Leak Detection 151
    Other Nonvolumetric Methods 153
Continuous In-Tank Monitors as Leak Detectors 153
Interpretation of Results 154
How to Use This Information to Reduce the Risk
   of a Bad Test 158

In-Tank Leak Detection Methodologies

H. Kendall Wilcox

The major objective of this chapter is to provide practical information for tank owners who have a vested interest in obtaining a valid test. In order to achieve this objective, the chapter covers the following subjects.

1. Introductory background, including general information about testing tanks.
2. Basic principles of tank testing, so that tank owners are aware of how the different tank tests operate. Volumetric, nonvolumetric, and continuous monitors are discussed separately.
3. Major factors affecting tank testing precision and accuracy, with emphasis on how to recognize a problem area.
4. Implications of these "noise" sources, which can obscure meaningful results.
5. Discussion of how tank owners can put the information to good use.

This chapter takes a generic approach to tank testing, in part to encourage the reader to think about the principles involved, but also because test developers are continually making changes in both hardware and methodology. It would be difficult to describe the current status of many methods accurately. In any case, it is not our purpose to promote or to denigrate any particular test method. It cannot be said which method is "the best"; all methods have their limitations and strengths. The choice of a method is determined by the test objectives, the specifics of the test location, economic considerations, availability of qualified test personnel, and many other factors.

This discussion is based on many field observations and some common sense. The reader should keep a few facts in mind, not only while reading this chapter, but also while reviewing the promotional claims of various vendors. For example:

- There are no perfect test methods.
- Some tests work better in some situations than others.
- The skill of the test operator is a key factor in obtaining a good test.
- Complexity should not be mistaken for quality. Sophisticated instru-

mentation such as computers and electronic "black boxes" do not erase the necessity for reasoned interpretation of the test data.
- Even the best operators occasionally make mistakes. Statistics indicate that it is highly improbable that any one approach will be 100% accurate each time it is applied.

BASIC PRINCIPLES OF TANK TESTING

The test methods discussed are limited primarily to those using equipment installed temporarily in the tank for test purposes only. These have been classified into either volumetric or nonvolumetric methods, based on how the tests are conducted and the type of data collected. Continuous monitors as applied to leak detection are also briefly discussed. Due to the complexity of the topic and the large number of methods now in existence, it is impossible to cover all of the methods and their individual differences in as much detail as one might like.

In short, volumetric methods are defined as those which produce a leak rate based on measuring the loss of product from the tank. Nonvolumetric methods are based on principles designed to detect a hole in the tank and do not depend upon a measured loss of product from the tank. In-tank monitors collect data continuously, and this information is usually combined with inventory reconciliation techniques to detect leaks.

There are currently a large number of methods, both volumetric and nonvolumetric, available for detecting leaks in underground tanks. A recent EPA report[1] describes 15 volumetric and 7 nonvolumetric test methods. The performance of several of these test methods was evaluated in a recent study conducted for EPA.[2] Since the preparation of these reports, several additional methods have been developed and marketed, and new ones appear regularly. At least 25 distinct methods are now in existence. Most of the volumetric methods are based on the same physical principles, but there are major variations in the hardware used to make the measurements. Nonvolumetric methods utilize a variety of measurement principles.

Unfortunately, marketers of most test methods are prone to making detection claims unsupported by test data. While the claims may be true in part, they are usually based only on the most accurately measured parameter (such as the capability to measure product level changes) but neglect the effects of other variables of larger uncertainty, which are always present. Since the total uncertainty of the test result is a function of all of the variables, including those which cannot be directly measured, the actual performance of a test may be substantially different from the promotional claims. Occasionally, a test is based on measurement techniques which are subject to so many uncertainties that the error rates must be unacceptably high. Fortunately, most such tests do not survive very long in the marketplace.

One of the problems faced by all test method developers is how to obtain data to confirm their test results. Comparisons between two methods are often made as if one method could invalidate the results of the other. Since all methods make at least occasional errors, the validity of this approach is questionable.

Confirming field data is a difficult task, particularly for small leaks. The careful removal and inspection of a tank, a time-consuming and expensive process, is probably the best way to verify test results. But the opportunity to do so usually presents itself only when tanks judged to be leaking are removed after testing. This verification process provides useful information regarding the frequency with which tight tanks are unnecessarily removed. Since very few tanks judged to be tight after testing are removed, however, the number of leaking tanks which remain in the ground is less well known. What should be apparent is that all testers will make a few errors, some more than others, because of the many variables inherent in testing tanks. The tank owner is well advised to keep this in mind when test results are marginal.

One of the problems associated with tank testing involves the definition of a "leak." The National Fire Protection Association (NFPA) has developed guidelines entitled "Recommended Practices for Underground Leakage of Flammable and Combustible Liquids."[3] A "precision test" is defined as "any test that takes into consideration the temperature coefficient of expansion of the product being tested as related to any temperature change during the test, and is capable of detecting a loss of 0.05 gal. (190 mL) per hour." Several other factors, such as tank distortions, must also be accounted for. The problem presented by this definition is that the conditions of a test are not well defined, and other factors in addition to those mentioned may adversely affect a test method's ability to detect a 0.05 gph loss. For example, the presence of a water table has a dramatic effect on the leak rate for some test methods and none at all for others. How the temperature changes are to be determined is also not defined in the NFPA guidelines. In addition, the size of the tank, which directly affects the ability of the tester to correct for temperature fluctuations, is not addressed. In fact, for tanks larger than 10,000 gal, it may be impossible to achieve the required 0.05 gph detection. Each of these considerations is discussed in more detail in the section on variables.

The result has been a proliferation of techniques, all purporting to be "precision tests," which give different leak rates when applied to the same tank under identical conditions. While the implications of this variance are not serious for large leaks, the situation becomes critical as the leak rate approaches the specified 0.05 gal criterion. If the leak rate is enhanced by one method (perhaps because the test calls for operating at a higher head pressure) and not by another, a tank may be indicated as leaking according to one method and certified as tight by the other.

It should be noted that the NFPA recommendations have frequently been incorporated into regulations, so that NFPA 329 has become a de facto standard for tank testing. In light of the many new developments and additional data, this document is currently being revised.

Volumetric Testing

Volumetric testing has the deceptive appearance of a simple, straightforward process. While this is true of the fundamental principles involved, a number of important variables often have a dramatic effect on the test results. Major sources of error are discussed later, after the description of volumetric testing.

Volumetric methods are based on principles which determine changes in the amount of product present in the tank. They characteristically give an estimate of the leak rate under the conditions of the test. Since conditions of the test may vary from method to method, different rates may be obtained from two methods when both are used on the same tank. In addition, variations in the environment (e.g., water table fluctuations) will affect the observed leak rate.

Although nearly all volumetric test methods are based on the same principles, many differ significantly in the approach to temperature compensation. Four of the better-known techniques include the classical approach, whereby temperature is directly measured; trend analysis, in which the change in level is measured at two head pressures; reference tubes, used to measure the effect of temperature changes; and self-compensating devices, which do not respond to level changes due only to temperature variations. These approaches are discussed in the following sections.

Not all test methods test the same parts of the tank system. Most methods are capable of testing at several product levels. In these cases, the product level can be raised until the entire tank system is under pressure for an initial test of both tank and piping. The level can then be lowered until the product level is below the piping, and a second test can be conducted. The results are then used to differentiate between leaks in the tank and leaks in the plumbing.

Some methods test only with the entire system under pressure. Leaks detected by this technique are reported as system leaks rather than tank leaks. No differentiation between tank and pipe leaks is possible with these techniques. Separate tests can be run on the delivery lines, but these separate methods do not test for the presence of leaks in lines other than delivery lines. Some methods are capable of testing only partially filled tanks. In these cases, the upper part of the tank and the plumbing are not tested.

Unfortunately, because of the varied approaches, there is no simple, all-encompassing description of volumetric testing. A typical test method, which illustrates most of the principles involved, is described here.

In a classical volumetric test, measurements of four experimental parameters must be conducted:

- measurement of the product level during the test
- measurement of the temperature changes in the product during the test
- determination of the coefficient of expansion of the product in the tank and the tank volume
- calibration of the test equipment under test conditions

A simplified schematic of a typical volumetric test apparatus is shown in Figure 1. The tank is almost always "overfilled" so that product is present in the fill pipe. This provides the most beneficial surface-to-volume ratio. Only a few methods have the sensitivity to test partially full tanks.

The classical volumetric test involves the simultaneous measurement of the product level and product temperature changes during a specific time interval. The level changes are converted into volume changes by a calibration factor, which is usually obtained by adding or removing known volumes of product from the tank. The temperature changes are converted into volume changes (ΔV_{Temp}) using the tank volume (V_{tank}), the coefficient of expansion (CE), and the ob-

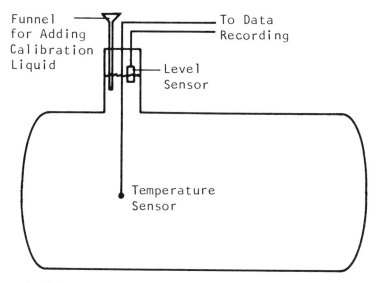

Figure 1. Schematic of a simplified tank test apparatus.

served temperature change of the product (ΔT) according to the equation:

$$\Delta V_{temp} = V_{tank} \times CE \times \Delta T$$

The volume change due to temperature change is then subtracted from the volume change obtained from the level measurements (ΔV_{Level}) to obtain a net temperature-corrected volume change (ΔV_{Net}) according to the equation:

$$\Delta V_{Net} = \Delta V_{Level} - \Delta V_{Temp}$$

If the volume change is due entirely to temperature effects, as would be the case for a tight tank, ΔV_{Level} and ΔV_{Temp} will be equal and the net value will be 0. If the tank is leaking, the volume changes due to level changes will exceed those due to temperature so that the net effect is that a loss of product is shown. It should be noted that several combinations of these two effects can be operating at the same time. Even though the level in the tank is observed to rise during the test, the test results may still correctly indicate a leak. In reporting the calculations, the usual convention is that losses are indicated by nega-

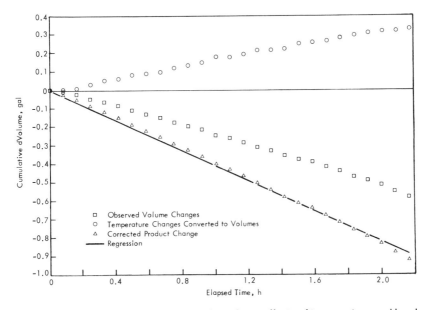

Figure 2. Typical test data illustrating the volume effects of temperature and level as related to the leak rate.

tive values. Thus, a leak of 0.05 gph is indicated as −0.05 gph leak rate.

Although it is not necessary to construct plots of the data to determine the presence of a leak, a plot of the data can be informative in interpreting the test data. A plot of a typical test where data were collected manually at 5-min intervals is shown in Figure 2. The temperature effects have been converted to volume and subtracted from the volume changes observed from the level changes to obtain a temperature-corrected leak rate.

Another technique used in several test methods has been called trend analysis. This approach uses the measurement of trends in the behavior of the tank at two different head pressures to detect a leak. The key measurement is the product level. Temperature measurements may be collected, but these do not figure directly into the calculation of the results. Any type of sensor can be used to measure the changes in level.

The underlying assumption is that the only variable affected by changes in head pressure is the leak itself. As the head pressure is increased or reduced, the leak rate will increase or decrease as well. If

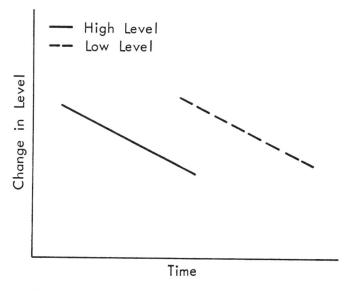

Figure 3. Typical trend behavior in a tight tank where the temperature is slowly decreasing. The slopes of the two lines are identical.

all other variables continue to behave in the same manner during the entire test, any changes in the behavior of the level at the two head pressures will be due to the presence of the leak. If no leak is present, the level behavior will be the same at both head pressures. If a leak is present, the rate will be higher at the higher head pressure, causing the rate of change in the level to increase. Schematic plots illustrating the behavior of a tight tank are shown in Figure 3; plots for a leaking tank are shown in Figure 4.

Temperature Compensation Techniques

Temperature is the single most important source of error in tank testing. All volumetric methods take temperature effects into account in some way. Although the measurement approach varies widely, test equipment and methods must accurately compensate for the average temperature behavior of the product.

Several approaches have been taken to eliminate the temperature variable. The classical approach, used by most methods, is to measure the temperature of the product directly by placing temperature monitoring equipment in the tank. Many methods rely on a single thermistor (or other temperature sensor) located at the vertical center of the

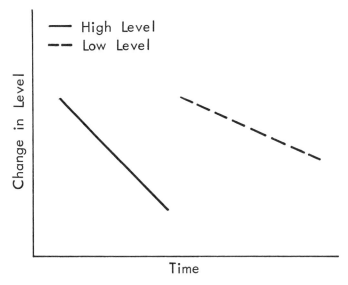

Figure 4. Typical trend behavior in a leaking tank where the temperature is slowly decreasing. The slopes of the lines are different.

tank. Other methods use an array of thermistors, ranging mostly from three to seven. The implicit assumption in using a single thermistor is that if stratification is present, the average rate of change will be represented by the central layer. The use of multiple sensors eliminates some of this uncertainty, because each thermistor represents a different layer. If, for example, temperature at the top of the tank is changing at a higher rate than in the rest of the tank, multiple sensors take this at least partially into account. The resolution of these devices is at least 0.01°F and is typically of the order of 0.001°F.

A key feature of temperature compensation techniques is the requirement that the tank be filled well in advance of the test so that temperature variations will have a chance to stabilize (the longer the stabilization time, the higher the probability of minimizing any variations). The minimum stabilization time required by most testers is around 4 hr, but it is recommended that the product be allowed to stabilize at least 8 hr or longer whenever possible.

An alternative approach to long stabilization times is to stir the product during the testing to achieve a uniform temperature throughout the tank. In this case, the outlet of the pump is located near the bottom of the tank, with the nozzle oriented to produce a swirling movement through the tank. At least one test method uses this approach. With this method, a single thermistor is located near the inlet to the circulation pump.

Three alternative approaches do not require direct temperature measurements to compensate for the effects of temperature changes. These are the use of reference tubes, self-compensating equipment, and trend analysis. These techniques are discussed below.

With the first approach, closed reference tubes are filled with product and placed in the tank. Since the tube is closed, the product level in the tube will respond to temperature changes but not to changes produced by leaks. The level in the tube is monitored at the same time as the level in the tank, and the temperature effects are subtracted from the results in the usual way. The major problem with this approach is that the sensitivity associated with the level in the reference tube is much less than that in the tank, because of different geometries. The reference tube is a vertical cylinder open at one end, while the tank is horizontal, with a small opening at the fill pipe. Since the surface-to-volume ratio is much more favorable for detecting changes in the tank than in the reference tube, temperature compensation using this approach can be seriously inadequate, unless the reference level can be measured extremely well. Use of very sensitive equipment, such as an interferometer, is required.

The second approach uses a float positioned in a partially filled tank so that level changes due to temperature are exactly offset by changes in density; that is, the position of the float is independent of temperature. The length of the float and the depth of placement are determined for each tank. In principle, the approach is sound; in practice, however, the application to horizontal tanks is limited to those that are only two-thirds to three-quarters full. For other levels, the error increases, because the float cannot be positioned properly in the tank. The limitation that the tank be only partially filled means that the upper quarter of the tank cannot be tested.

The third approach uses the measurement of trends in the behavior of the tank at two different head pressures to detect a leak, as discussed earlier. Since temperature is not affected by changes in level, this parameter does not appear in the calculation of the results. It is assumed that significant changes in the behavior of the temperature are not occurring during the testing and that the only parameter affected by the change in head pressure is the leak. What is important is not that other variables remain the same, but that if they are chang-

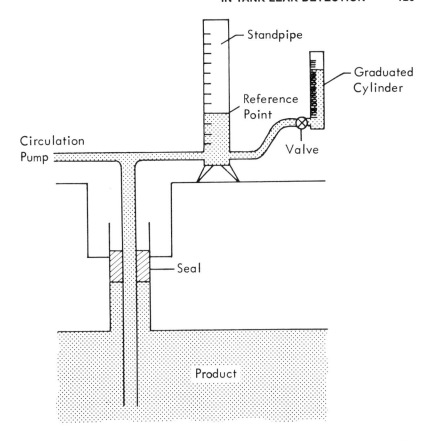

Figure 5. Schematic of device used to monitor changes in volume by maintaining product level at a constant height.

ing, they continue to change at the same rate during the entire test period.

Product Level Measurements

Product level measurements can be obtained through a wide variety of sensors, ranging from a simple ruler to a sophisticated laser interferometry system. Several sensors used in commercially available test methods are shown schematically in Figures 5-10. These diagrams are intended to show only the level monitor and do not illustrate a complete test apparatus. Figure 5 illustrates one of the simplest ways to monitor level changes; as the level changes during the test, the volume

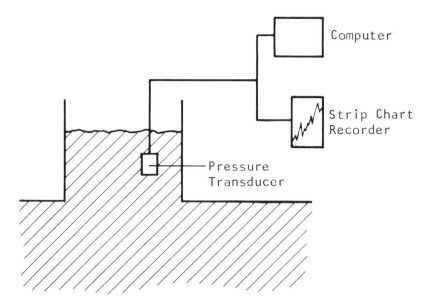

Figure 6. Level changes detected by changes in head pressure.

required to restore the level to the reference point is measured with a graduated cylinder. Figures 6 and 7 illustrate two ways in which pressure can be used to monitor level changes. In the first case, a sensitive pressure transducer is used to measure head pressure changes. The equipment is capable of measuring level changes as small as 0.001 in. under ideal conditions. The output of the transducer can be readily processed by computer or displayed on a strip chart. The transducer is usually placed close to the surface, so that level changes are large relative to the absolute head pressure.

Figure 7 shows a simple pressure-monitoring technique based on the pressure required to form bubbles from a tube submerged a short distance below the surface. As the surface level rises or falls, the pressure required to form the bubbles also rises and falls. The pressure can be recorded on a strip chart or monitored with a simple manometer, with the data recorded manually at periodic intervals. The resolution of this type of method is 0.005 in.

Figure 8 is a schematic of a level monitor based on changes in buoyancy produced on the probe by changes in fluid level. A sensitive balance or displacement sensor measures the weight losses of the buoyancy probe as the level fluctuates. The output of this device can

IN-TANK LEAK DETECTION   131

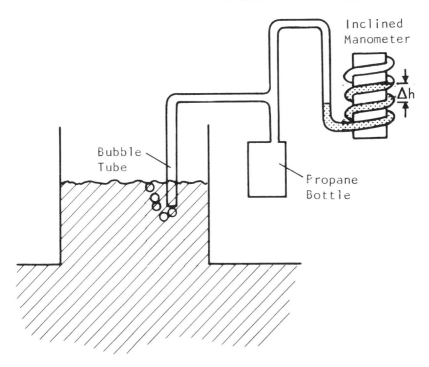

Figure 7. Level changes detected by change in pressure necessary to produce bubbles.

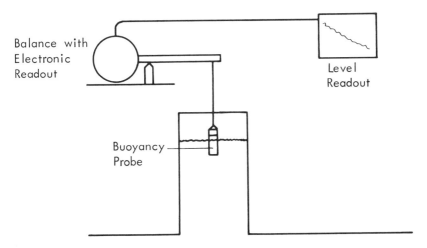

Figure 8. Use of a buoyancy probe to measure level changes.

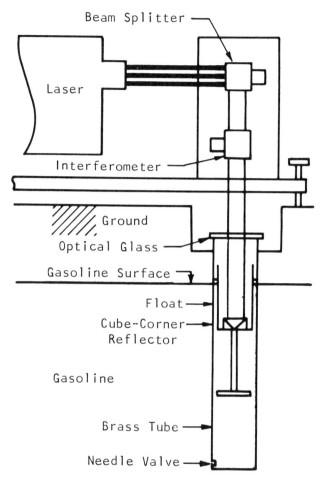

Figure 9. Laser interferometer used to measure level changes. (Source: U.S. EPA.[1])

also be either displayed on a strip chart or processed by a computer. Depending on the design, this approach can detect level changes of 0.0001 in.

The level monitor shown in Figure 9 is probably the most sensitive available. The measurement is based on a laser interferometer system with a resolution of one micro-inch (0.000001 in.). The equipment is relatively sophisticated and has been used primarily for research as a reference standard.

A method based on photometry is shown in Figure 10. The photo-

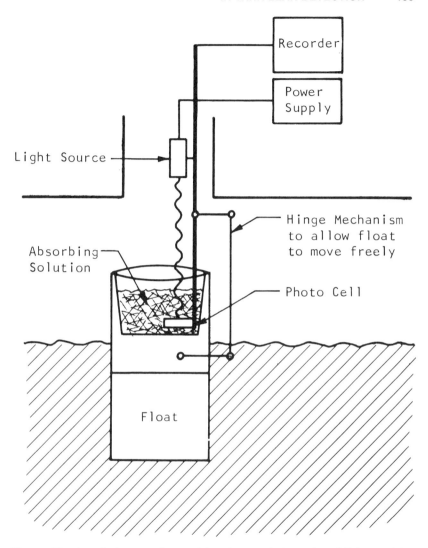

Figure 10. Level changes detected by changes in intensity of light reaching photocell. The depth of absorbing solution between the photocell and the light source changes as the float moves up or down.

cell and light source are affixed to the top of the tank. A float suspended below the photocell contains an ink cup. As the product level changes, the depth of the ink above the photocell changes, attenuating the intensity of light reaching the photocell. The technique is capable of detecting losses of 0.05 gph in a partially filled tank.

There are many other approaches, of course, but those shown here illustrate the variety of techniques used to monitor product level. It is generally true, especially for overfilled tanks, that product level measurements achieve the highest precision and, therefore, offer the least uncertainty of any of the factors affecting the test results. It is easy to detect changes in level corresponding to 0.01 gal or even less. This capability to measure level changes is usually cited in the promotional literature without acknowledgment that many other less precise and often unknown factors will reduce the accuracy and precision of the test method by a much larger amount.

To illustrate the importance of this parameter, two extreme cases can be considered: overfilled tanks, where the surface area is at a minimum; and partially filled tanks, where the surface area is at a maximum.

For an overfilled tank, where the only surface area open to the atmosphere is a 4 in. fill pipe, a volume change of 0.05 gal. corresponds to a level change of approximately 1 in. Since even the simplest method can achieve a measurement accuracy of 1/8 in., level changes corresponding to less than 0.01 gal. can be readily obtained.

On the other hand, for a half-filled 10,000 gal tank with a diameter of 8 ft and a length of 27 ft, a 0.05 gal change corresponds to a level change of approximately 0.00006 in. Only extremely sensitive level measurements, such as those obtained from the use of a laser interferometer, can detect such small changes. Of course, all variations of level are possible in a tank, but for most methods, as the product level drops below the top of the tank, the capability to detect small leaks decreases rapidly.

While overfilling to achieve maximum sensitivity is the usual practice, there are numerous situations where it is impractical, e.g., tanks containing hazardous materials, tanks with complicated plumbing, or situations where overfilling might produce a safety hazard. In addition, where there is a large leak in the plumbing, considerable product could be lost into the ground before the situation is recognized.

Calibration of Level Sensor

All volumetric methods rely on calibration of the test apparatus during the test. This must be done for each tank individually, since the surface area is a function of the number and size of openings (fill pipe, vent pipe, or other opening) in the tank. Calibration is accomplished through changing the product level in the tank by the addition

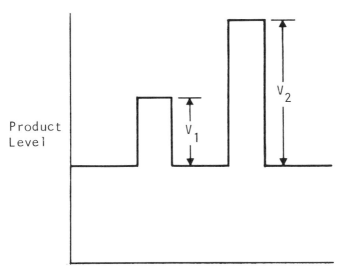

Figure 11. Product level behavior during calibration using two volumes.

or removal of a known amount of product or by inserting and withdrawing a rod of known volume into the tank. Figure 11, showing two different volume changes, schematically illustrates the behavior of product level during the process of inserting and removing calibration bars of volumes V_1 and V_2. The response of the test instrumentation can then be related to the known volume and a calibration coefficient determined. This process should be repeated several times and the results averaged.

As an example, if the addition of a bar of 0.05 gal volume produces a change in the level sensor of five divisions on a strip chart, each division then corresponds to 0.01 gal. A decrease of 15 divisions during a 2-hr test would then indicate a loss of 0.15 gal volume/2 hr or 0.075 gph.

Determination of the Coefficient of Expansion

The coefficient of expansion of the product in the tank must be obtained for each test so that the temperature effects on the volume of material in the tank can be determined. The relationship between the coefficient of expansion and the API gravity has been established by the American Petroleum Institute (API) and the results incorporated into two easy-to-use tables.[4]

Table 1. Typical Coefficients of Expansion for Selected Materials[a]

Gasoline	0.00065
Diesel Fuel	0.00045
Kerosene	0.00049
Water (at 68°F)	0.00011[b]
JP-4	0.00056

[a]Units are gal/°F
[b]Water varies substantially over the range of temperature normally encountered during testing and must be treated as a separate case.

The first step is to measure the API gravity and temperature of a sample of product obtained from the tank with a set of API hydrometers. The gravity is corrected back to a standard temperature of 60°F using API Table A. The corrected gravity is then used with API Table B to obtain the coefficient of expansion (CE). The CE is then used in the temperature correction calculations for the test.

Typical expansion coefficients are shown in Table 1.

VARIABLES IN VOLUMETRIC TESTING

Even though the equipment used on many tests is capable of producing a high degree of precision and accuracy in the measurement of both level and temperature, the problem of interpreting the data remains. Several important variables which cannot be readily measured directly can be present and can influence the test results. The most important of these are discussed in the following sections. It takes an experienced, skilled operator to recognize the subtle indications of these interferences and to compensate adequately for their presence. Even the best of testers can make an error with the result that a tight tank is pulled or a leaker is left in the ground. Either error could have a serious economic impact on the tank owner. Claims of "never having made an error" should be viewed with skepticism.

The major sources of uncertainty in the test results are these:

- temperature effects
- vapor pockets
- water table effects
- tank distortion
- vibration, including wind
- evaporation and condensation
- head pressure effects

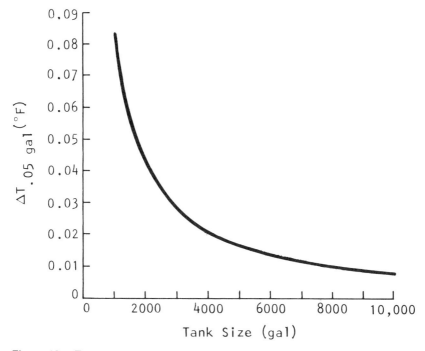

Figure 12. Temperature changes which will produce a volume change of 0.05 gal as a function of tank size.

Temperature Effects

Temperature has been recognized as the most important variable in volumetric testing. It may change in unpredictable and uncontrollable ways. Failure to measure or apply temperature corrections adequately can easily lead to incorrect conclusions (either false positives or false negatives).

While it is readily possible to measure temperature changes of the order of 0.001°F, being able to measure accurately is only part of the problem. The key question, difficult to answer, is whether or not the temperature changes monitored by the hardware adequately represent the true behavior of the temperature of the product.

Figure 12 depicts the temperature changes that will produce a volume change of 0.05 gal (the NFPA standard for precision tests) as a function of tank size. A change of only 0.01°F will result in a volume correction of 0.06 gal in a 10,000 gal tank containing gasoline. The problem of the temperature effect becomes even more severe with larger tanks, so it is questionable whether direct temperature compensation measurements can ever be made accurately enough to achieve detection of leaks in the 0.05 gal range.

Several approaches have been taken to eliminate this source of error. No one approach is perfect. Each has certain limitations and strengths. The classical measurement of temperature at a single location (or even at multiple locations) does not guarantee that the resulting measurement is representative; the temperature behavior in the tank may vary significantly from top to bottom. Usually the product is warmer at the top than at the bottom. There may, in fact, be several distinct strata as a result of surface heating effects or the delivery of product of a different temperature. Also, the effect of the ground temperature, the presence of a water table above the bottom of the tank, and thermal effects from the surface may all interact to produce differing effects on the strata within the tank. If these effects are not detected, the volume correction can produce significant errors in the test conclusions.

For example, suppose that the upper third of a 10,000 gal tank containing gasoline warms faster than the center and bottom segments by a rate of 0.01°F per hour. If not detected, this difference will produce an error in the volume correction of 0.02 gph. The use of multiple temperature sensors in this situation would greatly reduce the error.

Another potential source of error occurs when the tank is topped off just before testing. If the product added to the tank is of a different temperature than that in the tank, the equilibrium conditions of the tank may be upset. This is particularly true if a large amount of fuel is added, if the fuel is added near the temperature sensor, and if the temperature difference is large.

Thus, adequate stabilization times are important for methods which do not use stirring to achieve uniform temperature behavior. Although uneven temperature behavior is sometimes manifested in uneven temperature variations, a minimum stabilization time of 4 to 8 hr should be allowed. Overnight is preferred. Any indication of erratic temperature behavior of the product should be cause for allowing additional stabilization time.

Although there is some evidence that stirring is an effective way to produce uniform temperature behavior in gasoline,[2] the method is largely unevaluated with other liquids of higher viscosity, such as diesel fuel and fuel oils. Care should be taken to provide for adequate mixing time. It is possible in principle for heat to build up in the area around the pump outlet and inlet if the nozzle is improperly oriented or the material in the tank is viscous. This heat buildup would result in an overcorrection of the temperature effects, causing a bias toward falsely reporting a leak.

Instability in the temperature of the product is readily recognized from erratic temperature readings. Whenever this situation is observed, additional stirring is required. Testing should be continued until stable conditions are observed or enough data have been collected so that small short-term variations can be averaged out.

The relative sensitivity of reference tubes was discussed in general terms earlier and is covered in more detail here. Relative sensitivity can be calculated from the surface-to-volume ratio of the tank and the reference tube. To illustrate the magnitude of the problem associated with using reference tubes to determine temperature compensation, a 10,000-gal tank with only a 4-in. opening will exhibit a level change of approximately 1.2 in. if the temperature changes 0.01°F. At the same time, the level change in a reference tube 2 in. in diameter and 10 ft long containing about 1.6 gal will be less than 0.001 in. This means that the level in the reference tube is less sensitive to temperature change than the tank is by a factor of over 1,000. Unless the level in the reference tube can be measured very precisely, this approach will not give adequate temperature compensation.

A third alternative to the classical temperature measurement approach is using measurement of trends in the behavior of the tank at two different head pressures to detect a leak, as was discussed earlier. Since temperature is not affected by changes in product level, this parameter does not appear in the calculation of the results.

The only potential problem with this trend-analysis approach is the assumption that the behavior of the temperature during the test period is uniform. This is probably a safe assumption for most situations if adequate stabilization time has been allowed and the environmental conditions are not severe. As the length of the test period increases, the probability of a significant change in temperature also increases; this fact tends to shorten the time the tester is willing to allow for conducting the test. While shortening the test time may improve the situation with respect to temperature, it may cause other problems in interpreting results, because the effects of the leak are correspondingly reduced. The net result is a statistically uncertain test result. In addition, since there is a tendency to keep the test time short, testers may also not allow sufficient time between level changes

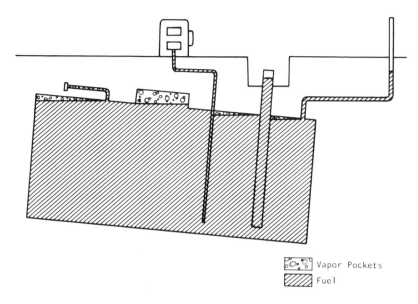

Figure 13. Schematic of the location of several vapor pockets in a tilted tank.

for tank distortions to equilibrate before they resume testing at the new level.

In summary, testers should be sensitive to erratic temperature behavior during tests which depend on direct temperature measurements. The use of multiple thermistors should be encouraged to verify the absence of, or to correct for, stratification effects. The use of reference tubes as a correction device should be discouraged for all methods except those employing adequately sensitive level monitors. Finally, when trend analysis is used as a technique in tank testing, the tendency to shorten test times should be carefully monitored.

Vapor Pockets

Vapor pockets are not generally a problem for test methods that do not overfill the tank. However, when a tank is overfilled, air and product vapors may be trapped at the top of the tank. They can be one of the most troublesome and expensive sources of error to eliminate from the test. Figure 13 shows a schematic of some of the common types of vapor pockets or bubbles which can be formed. It is difficult to recognize that bubbles are present, but most testers claim to be able to do so. The evidence is usually erratic behavior in the level measurements taken during the test. In some instances, the effects are visually observable. There has been no definitive study of this problem to date so the frequency and severity of the problem is not known. Test results which are otherwise unexplainable are often blamed on the presence of vapor pockets.

There are really only two options when vapor pockets are present in the tank: remove them, or use a test method that does not require overfilling to test the tank. It is unlikely that a valid test can be obtained if appreciable amounts of vapor are present. If a vapor pocket is present, it could mask a leak, or it could produce in a tight tank what appears to be a leak.

Removal of vapor can require the top of the tank to be exposed so that bleed valves can be installed. There are several points where vapor can be removed. If a manway is present, the trapped air pocket can be removed either by installing a bleed valve in the cover or by loosening the bolts to relieve the pressure.

A bleed valve may also need to be installed in the bung at the high end of the tank if the tank is tilted up from where the test equipment is located. Air trapped in the piping of the tank will cause the same problem as air trapped in the tank. Flanges which extend into the tank may also trap air along the top of the tank. This type of vapor pocket has been removed by inserting a J-shaped tube through the fill pipe and withdrawing the vapor with gentle suction. In some instances, a bleed valve may be installed directly in the tank. This should be attempted only by qualified contractors, as there is some risk of fire if it is not done properly.

Temperature and barometric pressure can have a significant effect on the volume occupied by a vapor pocket. If either changes during the test, the bubble will expand or contract, causing changes in the level of product in the fill pipe. The effect of a change in barometric pressure is shown in Figure 14 for three sizes of vapor pockets. As can be seen, the magnitude of the error for small bubbles is negligible for the barometric changes which might be encountered during a test. For large, 10- to 100-gal bubbles (which are not uncommon), however, the effect could be quite significant. Similarly, changes in temperature will also affect the size of the bubble, as indicated in Figure 15. Again, the effect is negligible for small bubbles or for small temperature changes.

A third factor that could cause a change in the size of the bubble over time is vaporization of product into the vapor space. Fuel will continue to volatilize and change the volume until the vapor space is

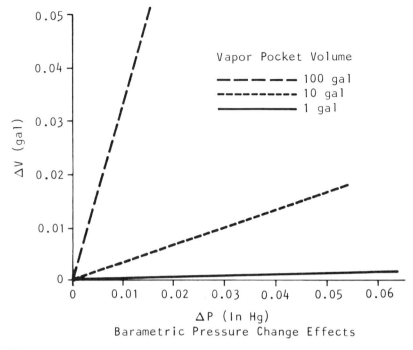

Figure 14. The effect of barometric pressure changes on the volume of a vapor pocket.

saturated. As the temperature or pressure changes during the test, the equilibrium of the vapor mix will shift, causing additional changes in the bubble volume until equilibrium is established. This process is beyond the scope of this discussion, and no experimental studies have been conducted that would relate directly to the testing of tanks. In the extreme case where the vapor pressure of the product is higher than the pressure placed on the tank (barometric pressure plus the fuel head), a bubble will form and continue to expand indefinitely.

One additional vapor pocket effect should be noted. If a large bubble is trapped where a hole is present, such as in a manway with a leak around the cover, loss of vapor through this leak will be large. The loss will continue until all of the vapor has escaped, whereupon the effect will suddenly disappear and the observed leak rate will become much smaller as liquid begins to flow through the hole.

As stated, vapor pockets can be a source of considerable difficulty and expense. If the temperature is rising and the barometric pressure

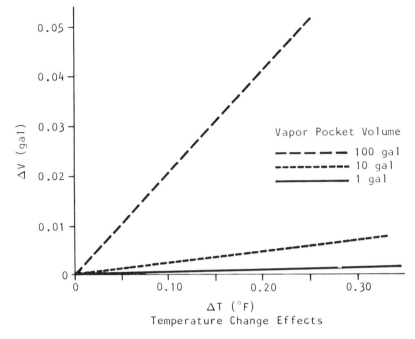

Figure 15. The effect of temperature changes on the volume of a vapor pocket.

is dropping (or vice versa), the combined effects on the size of the bubble may lead to an incorrect conclusion regarding the condition of the tank. The skill of the tester in detecting the vapor pocket is the best protection against error. Systems that collect data automatically, continuously, or at frequent intervals help in identifying patterns which characterize the presence of vapor. Erratic level data, particularly over a period of several hours, are a good indication of the presence of a vapor pocket.

In cases where it is impossible to remove the vapor (e.g., a tilted tank under a building), the best alternative could be to test using a method which does not require overfilling.

Water Table Effects

The presence of a water table above the bottom of the tank will have an effect on the loss of product through any holes below the water level, because liquids will flow through a hole only when pressure is applied. Since the presence of water will reduce the differential pressure between the inside and outside of the tank, the rate of liquid flow through any holes below the water table will be reduced. At some point the pressure will balance and no flow will occur either in or out. Since water is more dense than most petroleum products, the external pressure of the water will equal the internal pressure of the product when the water level is approximately 85% of the product level.

One way to compensate for the effect of the water table is to raise the product level during the test by extending the fill pipe above grade. At least one test method routinely applies this correction technique. When a water table is detected, the product level is raised until the product pressure at the bottom of the tank is 4 psi above the back pressure of the water table. This process produces additional pressure at the top of the tank, which may not be offset by the water table, so that leaks present in the top of the tank and in the piping are enhanced above what they would be if the water table were not present. In addition, this process exposes parts of the tank system that are not normally in contact with liquid, such as the vent lines, to product. Although this indicates a "leak" in the system, the leak is not environmentally significant unless the operator overfills the tank, either accidentally or by design.

Test methods that do not account for the water table run a considerable risk of missing leaks when a water table is present. In many instances, test methods do not even call for determining the presence of a water table, let alone taking active steps to compensate for its effect. At a minimum, every tank owner should require the tester, before testing, to determine the level of the water table, or ascertain that it is below the bottom of the tank. Steps can then be taken either to compensate for its effects or to abort the test before the full testing expenses have been incurred.

Tank Distortions

Distortions can occur in both steel and fiberglass tanks due to changes in the level of product in the tank. As the product level changes, the head pressure on the ends and walls of the tank also changes. After a period of time, which may vary from minutes to hours, the distortions are complete and the effect becomes insignificant. Distortions can easily produce changes in volume of over 10 gal.[3]

Tank distortions can sometimes be recognized because of their de-

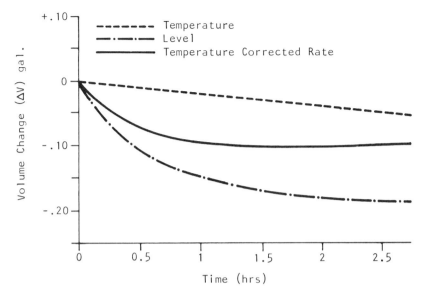

Figure 16. Behavior exhibited by the tank distortion which has not reached equilibrium.

creasing effect over time. An example of this type of behavior is shown in Figure 16. At first, as increased pressure is applied to the tank, the distortion occurs rapidly; then, as the position stabilizes, the process slows down until it asymptotically approaches equilibrium. In the example given, equilibrium was not established until about an hour after collection of test data began. Any test conclusions drawn during this first hour would have led to the conclusion that the tank was leaking. Data collected after this time period indicated that the tank was certifiable as tight. Test data exhibiting this type of behavior should be discarded and additional data collected until the effect of distortion is absent.

The time to reach equilibrium may vary with the type of backfill material, size of the tank, presence of a water table, etc. Although little information is available, indications from experienced tank testers are that fiberglass tanks may exhibit distortion effects for longer periods of time than steel tanks, so that additional stabilization time after changing the product level might be required before starting the test.

Test methods that require overfilling the tank should always consider this factor. The usual situation is for the tank to be filled to near the top several hours before testing to allow for temperature stabilization and deformation of the tank to occur. The tank is then topped off just before the test by adding a few gallons of product to the tank to bring the level of product into the fill pipe. Although the volume of product added may be small, the important factor is that the pressure on the ends and walls of the tank is increased substantially if the level is raised even a few inches. A change in level of 12 in. will change the pressure on the end of an 8-ft diameter tank by over a ton. Time allowances must be made for these effects to reach equilibrium each time the level is changed.

At least one test method routinely raises the initial level of product to the point at which the pressure at the bottom of the tank is 5 psi, thus forcing the deflection to occur rapidly. After some time, the pressure is reduced to 4 psi so that the tank can return quickly to its equilibrium position. This approach may have some advantages, in that the return to equilibrium is generally fast. Some limited data regarding this process were obtained during the EPA national survey of underground tanks.[5]

Test methods that rely on changing the head pressure on the tank to different levels should be particularly careful to monitor the data for evidence of tank distortion effects. Adequate time for the effects to reach equilibrium each time the level is changed should be provided before continuing the test.

A second source of tank distortion, observed many times, is the interaction of nearby tanks still in use. This can be particularly severe if the tanks are end-to-end. Distortions of the ends of an adjacent tank are transmitted through the backfill, producing distortion of the tank being tested. The effect can easily be in the range of 1 to 10 gal. The best solution is to remove all nearby tanks from service during the testing.

In summary, tank distortion effects can easily be overlooked, particularly if the test time is short. Tank owners should be sure that adequate precautions are taken by the tester to allow for tank distortions to be complete before testing.

Vibration

Vibrations can occur as a result of nearby traffic or windy conditions. Large vibrations produce ripples on the surface of the product that make accurate reading of the level difficult. The effect is most severe for methods that rely on measuring very small level changes.

To obtain valid test results, it is usually possible to extend test times, to delay testing, or even to remove the vibration source.

Evaporation and Condensation

Product loss due to evaporation is a potential problem, particularly during hot or windy conditions, where the tank is partially filled. Evaporation of product from the fill pipe could occur if the surface of the liquid is exposed during testing. If the test system is tight, no loss will occur from evaporation. One buoyancy method compensates for this effect by placing product in a cup which is part of the sensor mechanism. Evaporation losses from the fill pipe are then compensated for automatically by evaporation losses from the cup. Generally, condensation is a problem only if water condenses on the walls of the fill pipe and drains into the tank.

In general, evaporation and condensation do not seem to be a problem for most testers. Additional investigations would need to be conducted, however, to define the magnitude of this effect quantitatively.

Head Pressure Effects

Head pressure changes can have significant effects on the testing of tanks where a water table is present, or where the level of product changes during the testing due to temperature or some other factors. Head pressure effects have been exploited in the development of methods that depend upon trend analysis to detect leaks.

The effect is described mathematically by Torricelli's theorem, which is based on Bernoulli's principle, according to the equation:

$$v = \sqrt{2gh}$$

where v is the velocity of a liquid through an orifice, g is the gravitation constant, and h is the height of the liquid above the orifice. The equation states that the velocity of a liquid through a hole is related to the square root of the height, or head pressure, above the hole. A rigorous discussion of this principle can be found in most introductory physics texts. The theory applies only to small orifices with sharp edges and no back pressure, however; these conditions do not exist with underground tanks. But even though the calculated effects of head pressure changes will not be exact, they are nonetheless of use.

Torricelli's equation can be rearranged to a simple form for estimating the relative flow rates through a hole at different head pressures. The equation of interest is:

$$\frac{R_1}{R_2} = \sqrt{\frac{h_1}{h_1 + \Delta h_1}}$$

where Δh_1 is the change in product height, R_1 is the leak rate at head pressure h_1, and R_2 is the leak rate at head pressure ($h_1 + \Delta h$). If the leak rates and the change in head pressure are experimentally determined at the site during testing, the equation can then be solved in principle to determine the location of the leak. As stated, the results of the calculation will not be exact due to the effect of backfill, the water table (if present), and the nonuniform nature of the hole, but the information might still be of some use.

It is important to note that the effect of head pressure changes is the greatest when the hole is near the top of the tank, because the relative height change there is the largest. A plot of the effect is shown in Figure 17 for two sizes of hole. The practical implications of this are that the sensitivity of test methods using trend analysis depends on the position of the hole. For example, a leak of 0.10 gph located at the bottom of a tank 12 ft below the surface of the liquid will reduce to 0.09 gph if the product level is dropped to 10 ft. A leak of 0.10 gph located 3 ft below the surface of the liquid will reduce to 0.06 gph, a change that is larger by a factor of 4 than for the first case. Since trend analysis depends upon detection of these differences, it is obvious that very careful measurements or large head pressure changes will be required for the method to be successful.

Other volumetric methods will also demonstrate differences in their ability to detect a hole of a given size, depending on its location in the system. A hole resulting in a leak rate of 0.06 gph at a head pressure of 4 ft will leak at a rate of only 0.04 gph if the head pressure is lowered to 2 ft. The practical result is that a test method using a head pressure of 4 ft will not certify the tank as tight by current NFPA standards, but a test method operating at 2 ft will.

Water table effects will reduce the effective head pressure by the amount of back pressure present at the hole. Compensation for water table is usually based on the assumption that the hole is at the bottom of the tank. If a water table is present, and if the hole is above the water, the effect of compensating for the water table will be to en-

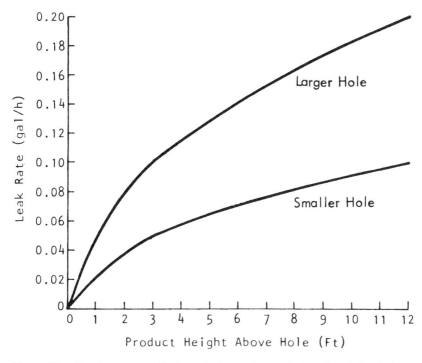

Figure 17. Head pressure effects on leak rate for two holes of undefined size.

hance the leak rate greatly. Because of these head pressure effects, it is well to test as far above the tank as possible. The owner should be aware of this phenomenon when evaluating the test data, particularly when trend analysis techniques are being used.

NONVOLUMETRIC TESTING

Nonvolumetric test methods are more varied than the volumetric and are based on principles that can detect a hole in the tank or piping. These methods may, in some cases, give an estimate of the leak rate, but they are not dependent on changes in the volume of product in the tank. Even though a quantitative leak rate is not obtained, some of these methods are extremely sensitive and are capable of detecting much smaller leaks than can most volumetric methods. Product losses of less than 0.01 gal can be reliably detected under some conditions by some methods. In fact, some of the techniques are almost too sensitive, in that they detect "holes" in the system through which little or no fuel will flow. Using these techniques as leak detectors will result in finding a leak in almost every system tested.

The attractiveness of nonvolumetric techniques stems from their insensitivity to many of the variables that cause difficulties with volumetric effects. For example, none of the methods is sensitive to temperature changes in the product in the tank, or to the presence of vapor pockets. This tends to make the data interpretation much less ambiguous, because there are fewer competing effects that can mimic or obscure a leak.

Nonvolumetric methods have not generally been recognized by regulatory agencies as acceptable test techniques, however, largely because they do not fit the NFPA's definition of a "precision test."[3] In addition, many claims by developers of these techniques have not been substantiated, either through extensive field use or by the independent evaluation of an unbiased third party. But this situation is apparently changing as more test experience is gained and the techniques are further refined.

Because of the attractiveness of many of the features of nonvolumetric techniques and the current high interest in tank testing, an increased awareness of this testing approach has developed. Several new test techniques have been developed and have received at least some field use. It is likely that this interest will continue, because the approach holds promise.

A few examples of measurement techniques that have received attention in the recent past will be described briefly. The discussion is not comprehensive, in part because definitive information on the performance characteristics of most of the techniques is not available. In addition, there is some overlap between the characteristics of some external monitors and nonvolumetric methods. The discussion here is limited to test methods that involve measurement techniques used on a temporary, short-term (one or two days at most) basis to ascertain the presence of a leak in a specific tank. The discussion excludes methods which require permanent or semipermanent installation, and methods which cannot distinguish which tank in a cluster is the source of an observed release.

Helium Leak Detection

All of the product is generally removed from the tank before a helium leak test is performed. Helium, which diffuses readily through soil and even concrete or asphalt, is then used to pressurize the tank.

The helium, which will diffuse rapidly through the leak and rise to the surface within minutes, is detected by an instrument based on mass spectrometric techniques.

In most cases it is helpful to drill small holes in the concrete or asphalt covering the tank. A matrix of holes is drilled for the initial evaluation. Additional holes are then drilled in areas where helium is detected. With experience, the tester can learn to make judgments on the size of the leak, as well as its approximate location.

Because the helium atom is small, it will diffuse through a hole approximately 50 times faster than will a gasoline molecule. Pipe joints that have not been properly tightened or coated with pipe sealer, as well as holes in the tank, will be detected as leaks. Vent lines, vapor recovery systems, manway covers, or any other location sealed when the tank was installed will be subject to severe testing even though these areas rarely, if ever, are in contact with liquid fuel. These commonly encountered situations virtually ensure that some small amount of helium will be detected outside the tank. The tester must then make a judgment call as to the severity of the problem and the necessity to excavate the tank.

The helium leak detection method is sometimes coupled with a differential pressure gauge intended to detect the loss of pressure over time. In principle the loss can be related to a leak rate, but in practice this is difficult to achieve reliably.

The helium leak test has not been widely used to test underground tanks because of the problems discussed above. The inconvenience of having to remove all of the product from the tank and the uncertainty in interpreting the test results in marginal cases have limited interest in the technique.

Tracer Leak Detection

The tracer leak detection method involves the mixing of an inert volatile tracer chemical with the fuel inside the underground tank. If product leaks out of the tank, the tracer chemical evaporates out of the product and diffuses into the air spaces in the backfill. Air is evacuated from the backfill at a high rate, typically 100 ft^3/min. Samples of the air are analyzed for tracer during the evacuation process. If tracer is detected in the backfill, a leak is indicated.

Figure 18 illustrates the fundamental process schematically. Also depicted in Figure 18 is a vapor tracer to aid in distinguishing between product leaks and vapor leaks. The vapor tracer is not mixed with the

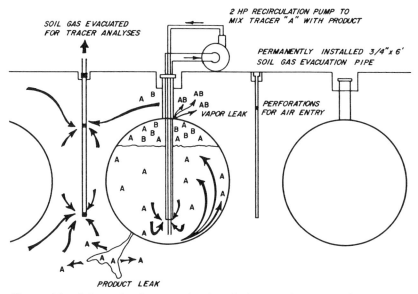

Figure 18. Schematic diagram showing dual tracer leak detection concept. (Source: report to Midwest Research Institute from Tracer Research Corporation.)

product, but is released into the air space above the product inside the tank. If a leak occurs above the product level in the tank, it will be indicated by detection of both the product tracer and the vapor tracer. If product leaks out, only the product tracer is detected.

The tracer leak detection method is made possible by the speed and extreme sensitivity of the tracer analytical method. The analytical method is a technique based on gas chromatography.

Measurements typically require 2 to 5 min, and the detection limits for tracer in the backfill air are in the low parts-per-trillion range. Product leaks as low as 0.001 gal/hr are easily detected with a tracer concentration in the product of 0.01% by weight. This tracer level represents a mixture of approximately 1 quart of tracer in 5,000 gal of product. This small quantity of tracer has no impact on the properties of the product.

The test technique is of interest because of its sensitivity and because it can be used without topping the tank or removing it from operation. This could be of considerable benefit in situations where use of the tank is critical. In addition, since there is no need to overfill or seal the tank for pressurization, preparation for the test is relatively simple even when the piping associated with the tank is complex. The method has been used on a variety of tanks and is now commercially available. Some application to large tanks (50,000 gal and larger) has been successfully demonstrated, a situation where volumetric tests are at their greatest disadvantage. If the test is conducted properly, the results are unambiguous, because the only source of tracer is within the tank.

Limitations of the method include the inability to test any portion of the tank below water. Additional work may also need to be conducted to determine its effectiveness in compact backfill, such as heavy clays.

Other Nonvolumetric Methods

Two other methods that should be mentioned include those based on the sound generated as product is forced through a hole or as air is drawn into the tank through a hole. Although both methods will in principle detect holes in tanks, the recognition of the sound patterns produced by the flow process can be difficult. Neither method has received widespread use in the field as a leak detection method for tanks, so data to evaluate their performance is very limited.

CONTINUOUS IN-TANK MONITORS AS LEAK DETECTORS

A large number of monitoring systems are currently available that can be permanently installed in a tank to collect information on product volume. Most are capable of measuring level changes of 0.1 in. and provide for some type of temperature compensation. They often incorporate a leak detection mode into the programming that will "test" the tank when it is not in use. The detection limit in the test mode is usually of the order of 0.2 gph, which is well beyond most regulatory requirements for tank testing.

Continuous monitors have seen little use as leak detectors, particularly for small leaks, because of the large uncertainty inherent in most of the measurements. These can easily range from 10 to 100 gal per reading and include delivery errors, meter errors in the retail pumps, and temperature uncertainties in the delivered and metered product. Data must be collected for a long period (weeks to months, depending on usage of the tank) before adequate data are available to reliably detect the presence of a small leak.

In addition, the process of extracting the information required to detect small leaks involves a fairly complex statistical analysis of the data, coupled with a good understanding of the operations at the site. While these processes have been worked out by a number of firms offering tank management services, the procedures are difficult enough that considerable training and effort would need to be expended before the typical tank operator could perform such an analysis. Some companies consider the procedures to be proprietary as well. In those cases, the data must be sent to another location for analysis.

In spite of these drawbacks, there are at least two benefits to continuous in-tank monitoring. First, if the equipment is already installed in the tank, the data necessary for the detection of leaks are already there. Second, if the method can be made competitive in sensitivity with periodic tank testing, the degree of protection from undetected losses is greatly enhanced. Since tank tightness testing is performed only periodically (almost never more often than yearly, and usually less frequently), the continuous monitor will provide better protection, even if several months of data are required for detection. Because the process is continual, leaks larger than the detection limit will not remain undetected for more than a few weeks. At that point, a more conventional tank test can be conducted to verify the presence or absence of a leak.

The statistical procedures required for this leak detection approach are beyond the scope of this chapter. Interested tank owners are advised to contact one of the tank management companies for more information on developing a leak detection program.

INTERPRETATION OF RESULTS

Treatments of the data after collection is one of the most important facets of tank testing. It does little good to collect high-quality data if the analysis techniques are inadequate. On the other hand, no amount of statistical massaging of the test results will turn bad data into good. One must know when to cut the losses and throw out the test.

Although the topic of data analysis is much too broad and complex to cover comprehensively in this chapter, some general observations are offered. Interested parties should locate a qualified statistician

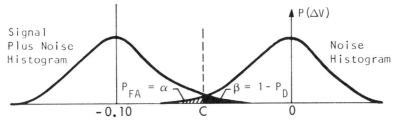

α = Probability of False Alarm, PFA
β = Probability of not detecting a leak of 0.05 gph, thus probability of detecting a leak of 0.05 gph is $1-\beta = P_D$
C = Cut Off for Declaring a Leak

Figure 19. Noise histograms for a nonleaking and leaking tank, showing probability of detecting false alarms and probability of detecting a leak of 0.10 gph when the criterion is set at 0.05 gph.

with experience in tank testing for a more detailed explanation of the statistical treatment of test data.

The objectives of this section are to illustrate some of the effects that the variables discussed earlier have on the data quality and to discuss the implications of these effects for determining a tank's condition—specifically, the frequency with which incorrect conclusions are reached.

Figure 19 is a histogram of the results of repeated testing of the same tank. The right-hand curve represents the spread of data that would be obtained from the repeated testing of a tight tank, while the curve on the left represents data on a similar tank with a leak of 0.10 gph. As the figure shows, if a tight tank is tested many times, most of the results will fall near 0. Due to random variation, the results will fall above or below 0 with decreasing frequency as the values move farther from 0. In a few cases, the observed value will fall below -0.05 gph, due purely to chance, and the tank will be declared as leaking even though it is actually tight. This frequency is designated as α in Figure 19.

Similarly, a tank leaking at 0.10 gph will show scatter in the results, with an occasional value falling above the -0.05 gph cutoff value (C). In this case, a leaking tank will be declared to be tight. This frequency is designated as β in Figure 19.

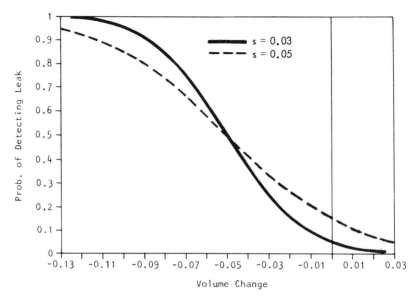

Figure 20. Probability of detecting a leak for C = 0.05 and values for S of 0.03 and 0.05.

Since the scatter in the data is related to the precision of the test, it is important to control the variables as tightly as possible, to reduce the number of erroneous conclusions. A very precise test will have less scatter in the data, with very few values falling into the overlap area. The cutoff value can be moved as well. For example, a value of C closer to 0 will result in detecting more leakers but will also result in the removal of more tight tanks. On the other hand, moving C toward larger leaks will result in the removal of fewer tight tanks at the expense of leaving more leakers in the ground. This is one case where you can't have it both ways at once.

An alternative way of presenting the same information is with a power curve, which can be used to determine the probability of detecting a leak of a given size when the cutoff value is defined as 0.05 gph. The shape of the curve depends on the standard deviation (S) of the test results, which is a measure of the precision of the test. Small standard deviations indicate higher precision.

Figure 20 shows power curves for two standard deviations, 0.03 and 0.05, which are within the noise range commonly observed for classical volumetric tests. The probability of detecting a leak is given on the y-axis for various leak rates for both values of S. Where the

IN-TANK LEAK DETECTION 157

true leak rate is 0.10 gph and S = 0.03, the probability of detecting the leak is approximately 95%.

This means that 5% of tanks with true leak rates of 0.10 gph will be declared to be tight. If the value of S is 0.05, the probability of detecting a leak of 0.10 gph falls to approximately 85% — so that 15% of the leakers will be declared tight. Similarly, a test with S = 0.03 will declare approximately 5% of the tight tanks to be leaking, while approximately 15% will be declared as leakers when S = 0.05.

Clearly, controlling the variables is extremely important, not only from the economic considerations of the owner, but also for the protection of the environment.

There are several reasonable ways to improve the precision of a test. One simple way is to extend the test time. This can be seen by considering the equation for computing standard deviation:

$$S = \sqrt{\frac{\Sigma (Xi)^2}{n - 1}}$$

where $\Sigma(Xi)^2$ is the sum of the deviation from the mean for each individual measurement, and n is the number of measurements.

As the equation shows, the value of S is reduced as the number of observations (n) increases. Longer tests or repeated tests will result in more observations, so that the probability for incorrect conclusions is reduced. This is the reason short tests should generally be avoided. There is simply not enough time to be certain whether the observed changes indicate a long-term trend due to a continuing leak or a short-term fluctuation due to the random effect of some variable not related to the leak. There are testers who will strongly argue the point, but a tank owner accepts a short test at his own peril. The cost of an incorrect conclusion about a tank's condition far outweigh the cost of additional test time.

A second way to reduce S is to gain better control of the variables that cause the large value. This would include better temperature measurement (perhaps by using additional temperature sensors), careful removal of vapor pockets, allowing for longer stabilization times after changing the product level, or addressing other factors related to a specific site. The corrective measure could be as simple as testing during the night, when there is less thermal effect from hot overburdens, less vibration from traffic, and less effect from the use of nearby tanks. Many testers have found the results of testing at night to be of higher quality.

A third factor to consider is in the selection of the test method. An evaluation of the test site can often reveal features that preclude the use of some types of tests. For example, if a tank is known to have trapped air pockets that are difficult to remove (e.g., the tank is under a building), the selection of a test method that requires overfilling is guaranteed to result in poor test quality. In this case, a method should be selected that can test a partially filled tank. A nonvolumetric method might also present an attractive alternative.

Last, proper treatment of the data can improve the quality of the results. A plot of the data such as in Figure 2 can be very helpful in interpreting the test results. Deviations due to tank distortion and erratic behavior of temperature or level become readily apparent when data points are visually displayed. It is difficult to recognize these patterns by inspection of tabular data. While a computer can make objective calculations, it cannot make subjective judgments about quality of data. The temperature-corrected leak rate (see Figure 2 for an example) can be obtained from the slope of the regression line

$$y = mx + b$$

where x is the time, y is the corrected volume change at time x, b is the y-intercept of the line, and m is the slope of the line. (The intercept value is usually set at 0.)

For tests where data are collected manually, the slope should be calculated from the data rather than by eyeballing a line. Most computerized systems do this automatically. The standard deviation of the slope should also be calculated for each test, so that its reliability can be estimated. The ways these calculations are performed are somewhat beyond the purpose of this chapter, but they are, in any case, the responsibility of the tester.

The combination of careful planning, appropriate test procedures, and proper treatment of the data will do much to protect the tank owner from poor test results. By being informed, a bad test will at least be recognized, so that additional testing can be conducted to resolve the issue.

HOW TO USE THIS INFORMATION TO REDUCE THE RISK OF A BAD TEST

After having looked at all of the potential problems associated with the detection of leaks in underground tanks, the tank owner may be tempted to conclude that any attempt to test tanks is doomed to failure. But this is not the case. Careful planning and insistence that testers follow certain practices can greatly reduce risk to the tank owner.

An important aspect of obtaining a valid test is the attitude of both the tester and the tank owner. Our observation is that many testers prefer the tank to pass the test. No one likes to be the bearer of bad news. This tendency is frequently exhibited by the efforts of a tester to collect "just a little" more data or to rearrange the test procedure slightly to get the test to "come in." Obviously, the owner would like to have the tank pass as well.

This attitude is risky, because the cost of leaving a leaker in the ground can be very disruptive to the bank account. Clean-up costs can run into the millions. The only prudent approach is to put maximum effort into determining the true condition of the tank, so that appropriate action can be taken. It is possible, under good conditions and by extending test times, to detect leaks smaller than the 0.05 gph limit set by many regulatory agencies. It would be foolhardy to ignore a leak just because it happened to fall below the legal limit.

At the same time, it is costly to remove a tight tank on the basis of imprecise test results. The probability of incorrectly declaring a leak when using an imprecise method (one where S is the same order of magnitude as the reported leak) is little better than that of tossing a coin. In these situations, a retest using a different approach or altering the test site to remove some of the noise sources is certainly in order.

Both of these situations illustrate the importance of approaching the test with the proper objective (to determine the true condition of the tank) whether the motivation is economic or altruistic.

A number of other factors should also be considered when obtaining a tester, and some of these are highlighted below. Admittedly, it is difficult to put some of them into practice. Nevertheless, we make the following recommendations for application before testing:

- Lots of bells and whistles, computers and other electronic gadgetry will not necessarily give good test results. Simpler is sometimes better. Don't be intimidated by complexity.
- Don't believe all the promoters' claims. With the high interest in the underground tank problem, new, unproven techniques are coming out of the woodwork. This can be a tough call. You shouldn't conclude that a method is not good just because it is new. Consider getting an expert opinion, if possible.
- Use only experienced, qualified testers. There are many charlatans out there who are only too eager to take your money. Consider that you rarely get more than you pay for. Quality may cost more in the short term but is far cheaper in the long term.
- Consider shutting down your entire operation when testing tanks, particularly if the tanks are close together. Interactive effects may be large enough to give misleading results. In addition, the effect of traffic is also reduced. If you are shut down at night anyway, this may be an excellent time to test.
- Consider the specific conditions at your site and try to select a method that will work to your advantage. For example, if your piping is complex, it may be very difficult to get the system ready to test for methods that require product levels above grade. Get help here too if you are uncertain. The EPA is currently conducting a series of evaluations of test methods; watch for this report (it will probably not appear until late 1987).

Once a test method has been selected, evaluate the method selected in light of the following recommendations:

- Insist that the tester determine the absence of a water table or its level, even if this is not the usual practice. You need this information to evaluate the test results. If the water table is high and the tester does not compensate for it, the test results may not mean much.
- Allow for as long a stabilization time as possible. Try to fill up a day ahead or overnight as a minimum.
- Try not to top off the tank immediately before testing. Get the level as high as possible before the test crew arrives, so that temperature and tank distortion effects have a chance to reach stable conditions. Some topping off is usually necessary, but try to keep the volume to a minimum, and allow time for equilibrium to be reestablished any time product is added or removed.
- Insist that testing be continued for a reasonably long time. At least 2 hr of data is desirable. If the test method is difficult to adapt to a single long test (as some of them are), ask for several repeat short tests. Don't allow the tester to get by on this one because he is

anxious to get to the next site and you are eager to reopen the operation. Long tests or repeated short tests are your best protection against bad test results.
- If the tester identifies vapor pockets as a problem, either get rid of them by whatever means (including excavating the top of the tank) or change to a method that does not require overfilling.
- In addition to the test report, get the raw data from the test as well. This is for your protection, and so that you can have an independent evaluation if you wish. You're the customer, you paid for the data, and you should be able to get it. Honest testers have nothing to fear.
- If the test results are marginal, consider a retest — if not immediately, then in a few months or a year. Don't make costly decisions based on questionable data.

We hope these suggestions and recommendations will be helpful to tank owners. While the situation is far from perfect, it is also far from hopeless. An informed tank owner has a high probability of obtaining a valid test if he keeps these various factors in mind.

REFERENCES

1. U.S. Environmental Protection Agency, "Underground Tank Leak Detection Methods: A State-of-the-Art Review" (EPA-600/2-86-001). Cincinnati, OH: U.S. Environmental Protection Agency.
2. U.S. Environmental Protection Agency, "Development of a Tank Test Method for a National Survey of Underground Storage Tanks" (EPA-560/5-86-014). Washington, DC: U.S. Environmental Protection Agency.
3. National Fire Protection Association, "Recommended Practices for Underground Leakage of Flammable and Combustible Liquids" (NFPA 329). Quincy, MA: National Fire Protection Association.
4. American Petroleum Institute, *Manual of Petroleum Measurement Standards*, Chapter 11.1 ("Volume Correction Factors"). Washington, DC: American Petroleum Institute, August 1980.
5. U.S. Environmental Protection Agency, "Underground Motor Fuel Storage Tanks: A National Survey" (EPA-560/5-86-013). Washington, DC: U.S. Environmental Protection Agency.

CHAPTER 5

Piping Release Detection and Monitoring

Todd G. Schwendeman

CHAPTER CONTENTS

Introduction 165
Pressure Monitoring 166
    Positive Pressure 166
    Negative Pressure 171
Tightness Testing 171
    Direct Testing 172
    Indirect Testing 172
External Sensing Systems 173
    Vapor Monitoring 174
    Cable Systems 174
Containment Technologies 174
    Low-Permeability Soils 176
    Impervious Barriers 176
    Double-Walled Piping 177
    Concrete Encasement and Soil Cements 181
Summary 181

Piping Release Detection and Monitoring

Todd G. Schwendeman

INTRODUCTION

The most integral part of an underground storage system is the piping or distribution system. The piping requires great care during installation and is exposed to a number of elements that can all contribute to a loss of the liquid contents. These elements include improper installation, vibration from the pumping system, surface loading forces (in traffic areas), temperature fluctuations, and ground movement (frost heaves, earthquakes, etc.). In addition, the number of pipe sections and fittings required to connect the piping are all locations that present the opportunity for a loss of system integrity.

Experience in the field has demonstrated that a greater percentage of leaks in underground storage systems occur in the piping system than in the tank itself. This has been documented by a number of states (California, Florida, and Maryland) and is considered common knowledge among tightness testing contractors.[1] Unfortunately, the piping is often overlooked in detection and monitoring programs.

Two pumping systems are commonly employed for distribution of substances from an underground storage tank: a remote or submerged pumping system, and a suction pumping system. The submerged system relies on a pump placed in or on top of the tank, and forces liquid from the storage tank to a dispenser under positive pressure. A suction pumping system relies on a suction pump located above ground, usually in the base of the liquid-dispensing device, which draws the liquid from the tank to the pump under negative pressure. In the last five years, remote pumping systems have gained popularity because of higher energy efficiencies, higher pumping rates, and the ability to service multiple dispensers with one pump. However, most existing installations have suction pumping systems. The type of pumping system used in an underground storage system will limit the types of leak detection and monitoring systems that can be effectively applied.

This chapter is designed to review the different technologies that can be applied to piping release detection and monitoring. A review of the different technologies includes pressure monitoring for both positive and negative pressured systems, tightness testing, external sensing systems utilizing both vapor monitoring and cable sensing applications, and, finally, containment technologies addressing impervious barriers and double-walled piping. An effective piping leak detection system may—depending on the degree of leak prevention necessary—incorporate several of these technologies into one detection and monitoring system.

PRESSURE MONITORING

Liquid will flow in a piping system when some form of pressure, positive or negative, is applied. Thus, the pressure under which a piping system operates is a readily monitored parameter to identify a loss of liquid. Several systems and techniques are available to monitor a piping system for losses.

Positive Pressure

Since the early 1970s, the use of remote or submerged pumping systems has become commonplace. API Publication 1635 states that "a line leak detector is recommended for all remote pumping systems."[2] However, the latest editions of NFPA 30 and the UFC Flammable and Combustible Liquids Code make no reference to line leak detectors. While pressure pumping systems offer a number of operational and maintenance advantages, a significant disadvantage is the high volume of liquid that can be lost through a perforation in the piping or a loose fitting while the system is under positive pressure. Large quantities of liquid can be lost from a positively pressurized line without any display of abnormal symptoms of performance.

A relatively simple procedure to test the integrity of a remote piping system is described in NFPA 329, Section 4-3.7.2.[3] The test procedure requires only the use of a 60-psi pressure gauge. The first step is to seal the dispensing line at the base of the dispenser and in the head of the pump. A gauge can usually be installed at the check valve in the head of the pump or the emergency shut-off valve at the base of the dispenser. The pump is then started to bring the line to its normal operating pressure (40-50 psi). This should seat the check valve, sealing the line. Any decrease in line pressure is then noted. If the pressure drops more than 5 psi/min, the valve should be checked and a retest should be performed to ensure that the pressure drop was not caused by compression of entrained air in the line. A pressure drop of over 5 psi on a retest indicates a possible leak in the line. Any pressure drop of less than 5 psi is inconclusive; the decrease may be a result of a small volume leak or temperature-induced liquid contraction.[3]

Although this is a useful test to determine tightness of a line on a periodic or spot-test basis, the method is not a substitute for, or intended to replace, more reliable continuous monitoring techniques. A number of different technologies can continuously monitor distribution piping under positive pressure. The technologies primarily utilize either electronic monitoring or mechanical monitoring.

The use of mechanical monitoring by means of a pressure-sensing diaphragm-operated valve is the most commonly employed method for monitoring underground motor fuel piping and has essentially become a petroleum industry standard. This reliable and relatively inexpensive ($150-$250) device has undergone a series of minor design refinements and is now a relatively tamper-resistant system.

The operation of the mechanical line pressure monitoring system is described below:[4]

The "Trip" or Relaxed Position. Under normal operating conditions it is assumed that the lines are filled with gasoline. When the system pressure is less than 1 psi, the diaphragm and poppet are in their "down" or "tripped" position. The position of the valve "poppet" is such as to allow approximately 1½ to 3 gpm to flow into the delivery line, through a bypass opening in the leak detector valve poppet, when the remote pump starts. Since the system is full, pressure builds rapidly and the poppet moves to the leak sensing position assuming there is no leak present. This is Position 1, as illustrated in Figure 1.

Leak Sensing Positions. As the pressure builds to approximately 8-10 psi, the diaphragm has moved the "poppet" to such a position as to almost stop the flow into the piping through the leak detector valve poppet. In this position, all the flow must then travel around the metering pin, and is limited to approximately 3 gpm. If a simultaneous loss from the system equals or exceeds this amount, the line pressure will not build beyond this point, and the valve will remain in the leak sensing position with the main flow blocked.

If there is an attempt to dispense while the valve is in this position, the line pressure will drop, the diaphragm will respond, and the poppet will return to Position 1, allowing only 1½ to 3 gpm to flow to the

UNDERGROUND STORAGE SYSTEMS

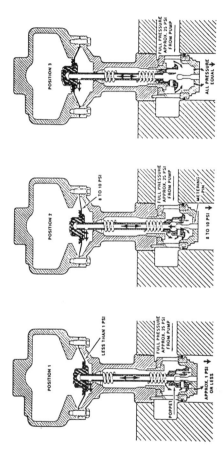

1. THE "TRIP" OR RELAXED POSITION. Under normal operating conditions, it is assumed that the lines are filled with gasoline. When the system pressure is less than 1 psi, the diaphragm and poppet are in their "down" or "tripped" position. The position of the valve "poppet" is such as to allow approximately 1½ to 3 gpm flow into the delivery line, through a bypass opening in the leak detector valve poppet, when the submersible pump starts. Since the system is full, pressure builds rapidly and the poppet moves to the leak sensing position assuming there is no leak present.

2. LEAK SENSING POSITION. As the pressure builds to approximately 8 to 10 psi (rapidly), the diaphragm has moved the "poppet" to such a position as to almost stop the flow into the piping through the leak detector valve poppet. In this position, all the flow must then travel around the metering pin which limits it to approximately 3 gph rate. If a simultaneous loss from the system equals or exceeds this amount, the line pressure will not build beyond this point and the valve will remain in the leak sensing position with the main flow blocked. If there is an attempt to dispense while the valve is in this position, the line pressure will drop, the diaphragm will respond, and the poppet will return to position 1 where 1½ to 3 gpm will flow to the dispensers. Leaks smaller than 3 gph will be indicated by the Leak Detector taking longer than two seconds to open completely. If there is no leakage in the system, the small flow around the metering pin increases the line pressure to 10 psi in approximately two seconds at which point the diaphragm will snap the poppet to position 3. This all takes place in less time than it takes to reset the dispenser, walk to the car, remove the gas tank cap, insert and open the nozzle.

3. NON-LEAK POSITION. This position allows full flow. The poppet will remain in this position as long as the system pressure remains above 1 psi. At less than 1 psi the poppet will return to position 1 and the next time the pump is activated, the Leak Detector will perform a line test.

Figure 1. Mechanical line leak detector. (Source: Red Jacket Pumps.)

dispensers. Leaks smaller than 3 gpm will be indicated if the mechanical leak detector takes longer than 2 sec to open completely. If there is no leakage in the system, the small flow around the metering pin increases the line pressure to 10 psi in approximately 2 sec, at which point the diaphragm will snap the poppet into Position 3. Position 2 is illustrated in Figure 1.

Non-Leak Position. This position allows full flow. The poppet will remain in this position as long as the system pressure remains above 1 psi. At less than 1 psi, the poppet will return to Position 1, and the next time the pump is activated the mechanical leak detector will perform a line test. Entrained air in the piping system, liquid thermal contraction, or a ballooning dispenser hose will not permit the leak monitoring system to operate properly and must be corrected.

A mechanical leak monitoring system is required for each pump used to dispense liquid from an underground storage system. The mechanical system will readily detect larger piping leaks (over 3 gpm), signaling that the leak is present by restricting the flow of liquid to a dispenser. For leaks smaller than 3 gpm the leak detector will open, but at a reduced rate relative to the size of the leak.

Electronic pressure monitors for piping systems are a relatively new technology. Even though there is only limited operational experience with these systems, the technology appears to have promise. Electronic leak detectors are more sophisticated than mechanical systems, are considerably more expensive ($2500–$6000 per facility), require a greater degree of maintenance, and have greater leak detection sensitivities. An electronic positive pressure monitoring system is illustrated in Figure 2.

One electronic system monitors a pressure decrease in the piping system over time. A system check is required when the pumps are activated at the opening of the dispensing system. After the pressurizing of each line, a check valve maintains 10–12 psi on the piping. The pressure in the line is monitored for 5 min. If the pressure drops below 4 psi, the pumps are reactivated, the lines are repressured, and the lines are monitored for an additional 5 min. If a similar pressure drop to below 4 psi occurs in the subsequent 5-min monitoring period, a leak is indicated and the system is shut down. An important function of this system is the inability of an operator to reactivate the system until the source of the pressure loss is corrected.[5]

Another electronic monitoring system is designed to monitor the rate of pressure increase in a piping system once the pumps are acti-

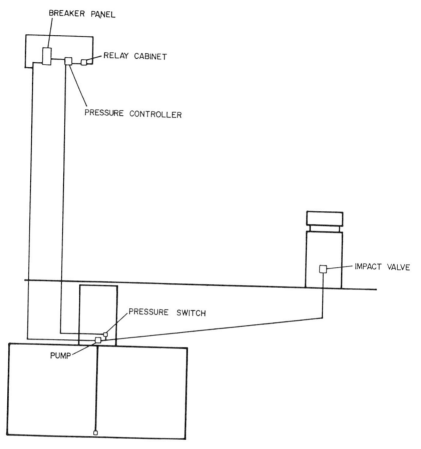

Figure 2. Electronic positive pressure monitoring system.

vated. For any given length of piping between the pump and the dispenser, the amount of time it should take for that length of piping to become fully pressurized can be calculated. This value is then programmed into the detection device. Should the pressure not rise quickly enough for the programmed length of piping, a leak will be indicated and the system will be shut down. This electronic monitoring system also prevents reactivation of the dispensing system once a release from the piping is identified.[6]

Negative Pressure

Pumping systems located above ground in the dispensers operate by using negative pressure (vacuum) to pull the liquid through the piping system. At present there are no continuous monitoring devices for detecting leaks from single-walled, suction system piping. However, by carefully observing the performance of the liquid dispensing device, the operator can detect the symptoms of possible line leaks. If a leak is present in a suction pumping system, air will typically enter the pipe as liquid drains back into the tank through the check valve (depending upon its location) or into the surrounding environment. Air in the piping will be readily identifiable by the action of the pump in the first few seconds of operation. The characteristics of entrained air include:[7]

- Display wheels skip or jump before liquid is dispensed.
- The pump runs, but does not pump liquid.
- The pump seems to overspeed when first turned on and then slow down as it begins to pump liquid.
- There is a rattling sound in the pump and erratic liquid flow, indicating that air and liquid are being mixed.

Under any of these conditions, the initial step to verify a leak should be to inspect the check valve. Check valves can be located close to the pump inlet, mounted in the piping above the tank, or on the end of the suction stub inside the tank. If the check valve seals tightly, and any of the above symptoms still is present, leak testing of the piping should be performed.

TIGHTNESS TESTING

Underground storage system tightness testing is a rapidly expanding field marked by the application of many innovative technologies. This chapter does not attempt to describe all forms of tightness testing comprehensively; the discussion is restricted to the most commonly used piping tightness testing methodologies that adhere to National Fire Protection Association (NFPA) 329 criteria. For discussion purposes the methodologies have been divided into two categories: direct testing and indirect testing of an underground piping system.

There are several philosophies on how and when tightness testing should be applied. A number of state and local regulatory agencies have mandated periodic tightness testing as an acceptable monitoring or leak detection option. However, the American Petroleum Institute recommends that tightness testing be used only to investigate or confirm that a leak has occurred.[7,8]

Direct Testing

Direct tightness testing can be defined as any form of tightness testing that tests only the piping system for integrity. These systems typically use a type of operating principle or technology similar to that of the tank tightness tester, yet are conducted independently of the test on the tank itself.

Hydraulic testing of the lines is the most commonly used piping system tightness testing methodology. A hydraulic tester is connected by an adapter to either a suction pump or a remote dispenser. Entrained air is removed from the piping system by means of a magnetically attached automatic air bleed. Hydraulic pressure, equivalent to approximately 15 psi for suction piping and a minimum of 50 psi for remote pumping systems, is applied by a hand pump. The pressure is applied to the piping leading back to the angle check at a tank or the remote pump.

If pressure is lost from the piping system, the necessary liquid (measured in increments of 0.001 gallons or 2.5 mL) is reintroduced to restore the test pressure. Measurement of a leak is recorded according to volume/time lost.

Indirect Testing

Indirect testing can be defined as tightness testing of the piping system as a component of a full system test (not a separate test). Indirect tightness testing of the piping system involves the application of a full system tightness test that is performed on an underground storage system completely filled with product. The test is conducted twice: once for both the tank and the piping, and the second time for just the tank. The integrity of the piping system can be determined based on the results. The many different tightness testing technologies that are discussed in Chapter 4 are also applied to indirect piping testing. These include buoyancy, hydrostatic head, ultrasonic, and float systems.

An indirect test is typically performed by raising the liquid level to the point of the highest unexposed piping. Where necessary, risers are used for this purpose. If a release is detected during this full system test, the level is lowered to the top of the tank and another test is conducted. If the system is tight at the top of the tank, it is concluded that a piping leak exists; however, should the system still be leaking during the tank top test, the liquid level is typically lowered to below the top of the tank and another test performed. Thus, through a process of elimination, the level at which the leak is occurring can be determined.

Indirect testing can require additional time to eliminate the various level testing options when a piping test is conducted. In addition, a conclusive final full system test should be performed after any necessary repairs are made, to ensure that the entire system, tank and piping, is tight.

EXTERNAL SENSING SYSTEMS

External sensing systems are piping leak monitoring or detection systems that monitor the environment outside of the piping system. These systems are typically liquid or vapor sensors, technologically similar to the devices for monitoring a tank excavation that are described in Chapter 3. The systems function most effectively when deployed within the piping trench backfill and relatively close to the piping system.

To date, external sensing systems to monitor underground piping for leaks have received limited application. Most of the technologies are relatively new and expensive and lack field operational experience. Furthermore, there are no standards that stipulate sensor placement, operational guidelines, or the number of sampling locations for a given length of piping.

External sensing systems that monitor the groundwater table, the most common type of external sensing system, are not suited for piping applications, since the groundwater is rarely located as close to the ground surface as the piping. Thus, a loss of liquid would affect a significant amount of soil before reaching the water table and triggering an alarm. These systems fail to provide early warning of a piping release unless an impervious barrier or collection basin, which permits any lost material to accumulate, is placed beneath the piping.

The two external sensing systems that have seen limited application to piping system monitoring are vapor monitoring techniques and liquid detection cable systems. Only the vapor monitoring methodologies can be used at existing facilities; a liquid detection cable system requires installation beneath the piping and is therefore impractical for existing systems, except when the system is used in conjunction with a sump.

Vapor Monitoring

Vapor monitoring in the piping system excavation is a promising technique for piping leak monitoring. However, for this technology to become viable, interferences from surface spills, overfills, and other sources will have to be eliminated, and the number of monitoring locations per length of piping for various sensing technologies must be established. Vadose zone monitoring is gaining acceptance, and more operational experience for organic liquids release monitoring is being acquired. The different types of methodologies available to monitor vapors in the vicinity of the piping system excavation are similar to the methodologies used to monitor a tank excavation with reduced sampling depth. Figure 3 illustrates an application of vapor monitoring for a piping system.

Cable Systems

A piping system can also be monitored through the use of a cable system. A cable system consists of a continuous cable (electronic or fiber optic) laid beneath the piping in the piping trench and connected to a controller or an alarm panel. The cables typically incorporate some type of hydrocarbon-degradable material, styrene butadiene copolymer, which triggers an alarm on contact with a released substance. Degradation of the styrene butadiene copolymer material results in a change in electrical conductivity or light transport, thus signaling a release.

Since a cable system must underlie a piping system, this approach is feasible only for new systems. In addition, the hydrocarbon-degradable materials are not regenerative; when they come in contact with hydrocarbon, they must be excavated and replaced. Costs for a cable monitoring system range from $3 to $7/ft, not including installation.

PIPING LEAK DETECTION 175

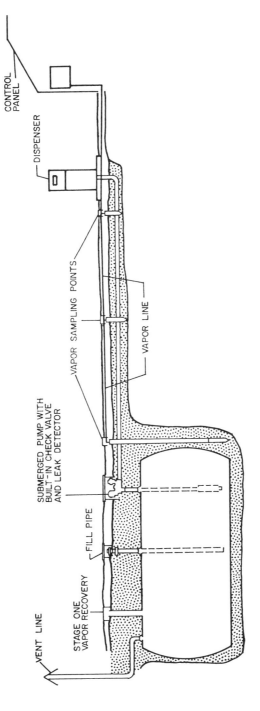

Figure 3. Vapor monitoring of a piping system.

CONTAINMENT TECHNOLOGIES

Containment technologies were developed to reduce the contact of a leaked substance with the environment. Secondary containment, as those technologies are commonly called, is accomplished by establishing a barrier around the piping system so that a leak or loss of material will not migrate into the environment. Five technologies have been applied as piping secondary containment: low-permeability soils, impervious barriers, double-walled pipe, concrete encased piping, and soil cements.

Low-Permeability Soils

Low-permeability soils (soils having a permeability less than 10^{-6} cm/sec) have been used as a form of secondary containment for the piping system. Typically, the use of low-permeability soils has been restricted to clay soils that occur naturally onsite. Only on rare occasions have low-permeability soils been transported to a site and applied to piping trenches.

Lining the piping trench with low-permeability soils is difficult. The material is usually hard to handle, is expensive to transport, and requires compacting in the piping trench and quality control testing during the installation. Once the installation is completed, the final cost is usually greater than other forms of piping secondary containment.

Impervious Barriers

The typical impervious barrier applied as secondary containment to underground piping systems is a flexible membrane liner. The liners are thermoplastic or polymeric sheets, typically 50 mm thick, that are designed to be impervious to the stored substance, resistant to chemical or biological degradation and climatic influences, and capable of withstanding limited structural stresses. Geotextile materials can be placed in the piping excavation to improve the load-bearing capacity of the liner and to reduce liner chafing. To date, no long-term reliability data are available on the durability of flexible membrane liners in a field application.

Flexible membrane liners are generally produced by the plastics and rubber industries and range from vulcanizable to nonvulcanizable (thermoplastic) plastics and rubbers. These materials vary in chemical polarity and crystallinity, and they can be manufactured with different fillers, antidegradants, and plasticizers. Thus, commercial products based on the same polymer can vary from manufacturer to manufacturer.[9]

Materials for flexible membrane liners specific to hydrocarbon usage include polyester elastomer, high density polyethylene, epichlorhydrin, and polyurethane-based products. The cost for the liner materials ranges from $0.22 to $2.50/ft^2.[8]

An important variable in the selection of a flexible membrane liner is the ease of installation for an underground piping system. A considerable amount of the cost of a piping trench liner is directly related to installation. Seam welding and the piecing together of difficult-access areas (beneath buildings, beneath dispensers, etc.) will require additional installation time.

A flexible membrane liner for the piping system can be used in conjunction with a lined tank excavation as well as double-walled tanks. Figures 4 and 5 illustrate these two scenarios. When a liner is used in the piping trench, the trench is typically sloped away from the tank excavation to differentiate piping failures from tank failures. A sump is located at the lowest point of the piping system to facilitate monitoring at this point. Any of the monitoring systems described in Chapter 3 can be effectively used in the sump. Flexible membrane liners can also be used as a tube to encase the piping. When a tube is formed, the seam should run along the top of the membrane.

As the use of flexible membrane liners has increased, many of the complexities involving the initial installations have been eliminated. Because an increasing number of installation contractors are becoming familiar with the handling and assembly of these systems, costs should be reduced and performance improved. Some difficulties remain regarding the testing of a flexible membrane system for tightness and in conducting repairs.

Double-Walled Piping

Double-walled piping has only recently been introduced onto the market but is becoming increasingly available. The concept is much the same as for double-walled tanks, and consists of a primary container (inner pipe), a secondary container (outer pipe), and an interstitial space that can be monitored for losses. A double-walled retail petroleum facility piping system typically consists of standard 2-in. (i.d.) piping, constructed from fiberglass-reinforced plastic (FRP)

178 UNDERGROUND STORAGE SYSTEMS

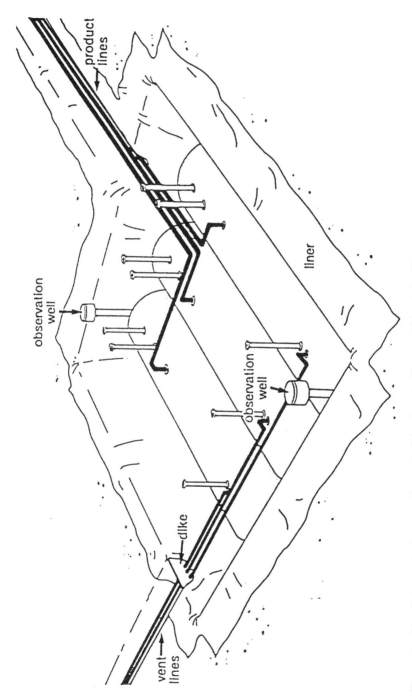

Figure 4. Lined storage tank excavation and piping trench. (Source: Eklund et al.)[9]

PIPING LEAK DETECTION 179

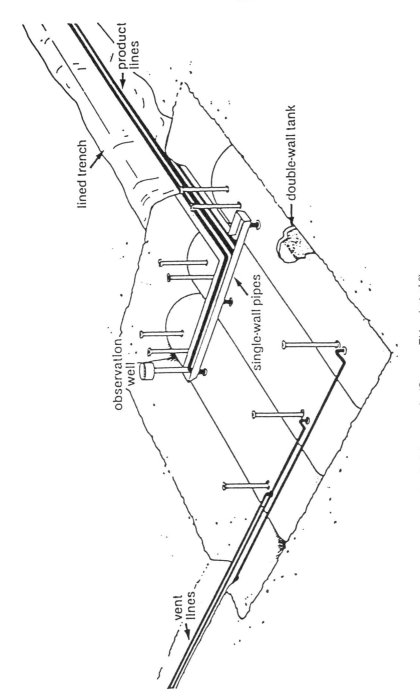

Figure 5. Double-walled tanks with a lined piping trench. (Source: Eklund et al.[9])

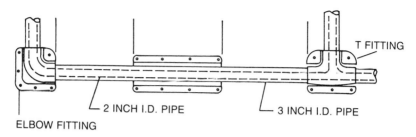

Figure 6. Double-walled pipe and fittings.

and designed to handle the liquid of concern, inside the next larger size pipe, usually 3-in. (i.d.) piping. The outer piping fittings are custom-fabricated and joined with fasteners.[10] Figure 6 illustrates double-walled pipe and fittings.

For a double-walled piping system to be effective, care must be exercised during handling and installation. FRP double-walled piping is relatively brittle and requires close inspection during installation to assure the integrity of the system. A trained installer should be employed to ensure a high-quality, leak-free double-walled piping system.

Testing a double-walled piping system requires that the inner pipe be tested before the outer pipe is assembled. The inner pipe should be tested at 50 psi or 1.5 times the working pressure of the system; however, the outer piping should be tested only at 5 psi.

A double-walled piping system can be monitored either passively or actively. The most common monitoring approach has been to slope the piping away from the tank excavation to some form of large containment structure or vault. Depending on regulatory requirements, the vault can be periodically visually monitored, or a continuous monitor system, as described in Chapter 3, can be used.

Another approach that has been applied is the spacing of small sumps along the length of the piping. These strategic access points allow sections of piping to be monitored (again by means of systems as described in Chapter 3) so that the specific section that may have a problem can be isolated. When sumps or similar technologies are not used, it may not be possible to determine where the leak may be in the piping run.

In addition, double-walled pipe can be monitored with a hydrocarbon-sensitive cable. The cable is highly sensitive to hydrocarbon contact and the location of the leak can be pinpointed. Should a leak be detected, the section of the cable which came into contact with the released material has to be replaced.

Double-walled piping materials cost approximately $10–$12/ft. Installed, a double-walled piping system costs two to three times as much as a single-walled system.[11] Until the use of double-walled pipe gains greater acceptance, only limited information concerning the field reliability of the system is available.

Concrete Encasement and Soil Cements

Concrete encasement has, in limited applications, been used as a form of secondary containment for piping systems. Once the piping is installed, concrete is used for backfilling the trench instead of sand or pea gravel. Concrete encasement provides only a temporary impervious barrier, however, as over time the concrete typically fractures because of backfill movement and loading stresses. In addition, piping repair and maintenance is laborious and expensive. Concrete encasement is an impractical form of secondary containment for most piping installations.

Soil cements have also been used to a limited degree as a form of secondary containment for piping. The soil cement, typically portland cement or bentonite, is mixed with native soil in the piping trench before the piping is installed. Once the mixture has cured (24–72 hr), the piping can be installed and the trench backfilled. Because of difficulties in attaining a fully impervious barrier, and because of the curing time of the material, this technique also has limited application as an approach to secondary containment.

SUMMARY

The piping system is the most vulnerable component of subsurface liquid storage systems and is subject to the greatest operational stresses and external influences of the storage system. Considerable care is required during a piping system installation to assure that the system is leak free. A line leak detector for positive pressure piping is a common monitoring requirement in those states that have developed underground storage regulatory programs.

Even though a significant proportion of releases from an underground storage system occur in the piping, there are only a limited number of practical, cost-effective monitoring options. The technique that has received the greatest application is mechanical monitoring of positive operating pressure in the piping system. Mechanical monitoring is relatively tamper-resistant, easy to install, and inexpensive. Electronics is also being applied to monitor positive operating pressure. Negative pressure (suction pumping systems) is best monitored by pumping characteristics that manifest themselves during dispensing (meter spin, rattling sounds, entrained air, etc.). External sensing systems that use vapor monitoring hold promise; however, cable systems are limited to new installations and are difficult to replace. Groundwater monitoring of piping systems is not feasible in most situations.

Piping system tightness testing is another approach to determine the integrity of the system. Direct tightness testing is performed upon the piping independent of any tank tightness testing; indirect tightness testing is conducted in conjunction with a tank tightness test. Indirect testing is typically accomplished by lowering the liquid level in the tank below the elevation of the piping to conduct a retest.

Containment technologies are being used in highly sensitive environmental locations and have been mandated in certain states for all new installations. An impervious barrier system consists of an impermeable, flexible membrane that either lines the bottom of the piping trench or surrounds the piping. Double-walled piping is a pipe within a pipe and permits monitoring of the interstitial space. Double-walled piping has limited field experience and requires specialized installation. Both containment systems can be used only with a new installation. Concrete encasement and soil cements have received only highly limited use.

Substantial research should be applied to develop a cost-effective, low-maintenance piping monitoring system that can be applied to both new and existing installations.

REFERENCES

1. Personal communication with George Lomax, Heath Consultants, Stoughton, MA, March 1986.
2. American Petroleum Institute, "Recommended Practice for Underground Petroleum Product Storage Systems of Marketing and Distribution Facilities" (Publication 1635). Washington, DC: American Petroleum Institute.

3. National Fire Protection Association, "Underground Leakage of Flammable and Combustible Liquids" (NFPA 329). Quincy, MA: National Fire Protection Association, 1983.
4. The Marley Pump Company, "Red Jacket Leak Detector Operating Instructions." Mission, KS: The Marley Pump Company.
5. K and E Associates, Petroleum Monitoring System (PMS-800). Long Beach, CA: K and E Associates, 1986.
6. Ronan Engineering Company, Leak Detector System Model TRS-76. Woodland Hills, CA: Ronan Engineering Company, 1986.
7. American Petroleum Institute, "Recommended Practices for Bulk Liquid Stock Control at Retail Outlets," third edition (Publication 1621). Washington, DC: American Petroleum Institute, 1977.
8. American Petroleum Institute, "Recommended Practice for Underground Petroleum Product Storage Systems at Marketing and Distribution Facilities" (Publication 1635). Washington, DC: American Petroleum Institute, December 1985.
9. Eklund, A. G.; Dickerman, J. C.; and Harris, H. E., "Secondary Containment For Underground Petroleum Products Storage Systems at Retail Outlets: Marketing Department Research Report." Washington, DC: American Petroleum Institute, September 1984.
10. Ameron Piping, "Secondary Containment System: Installation Instructions." Houston, TX: Ameron Fiberglass Pipe Division.
11. Personal conversation with Reed Van Cleave, sales manager, Petroleum Marketing, Ameron Fiberglass Pipe Division, Houston, TX.

APPENDIX A

Hazardous Substance List, Comprehensive Environmental Response, Compensation and Liability Act of 1980, Section 101(14)

(For regulation of underground storage tanks under the Resource Conservation and Recovery Act, Subtitle I. This list of CERCLA hazardous substances was published in the *Federal Register* on April 4, 1985 [50 FR 13546].)

Hazardous Substance	CASRN[a]	Hazardous Substance	CASRN
Acenaphthene	83329	Acetonitrile	75058
Acenaphthylene	208968	3-(alpha-Acetonylbenzyl)-	
Acetaldehyde	75070	4-hydroxycoumann and	
Acetaldehyde, chloro-	107200	salts	81812
Acetaldehyde, trichloro-	75876	Acetophenone	98862
Acetamide, N-(aminothioxomethyl)-	591082	2-Acetylaminofluorene	53963
		Acetyl bromide	506967
Acetamide, N-(4-ethoxyphenyl)-	62442	Acetyl chloride	75365
		1-Acetyl-2-thiourea	591082
Acetamide, N-9H-fluoren-2-yl	53963	Acrolein	107028
		Acrylamide	79061
Acetamide, 2-fluoro-	640197	Acrylic acid	79107
Acetic acid	64197	Acrylonitrile	107131
Acetic acid, ethyl ester	141786	Adipic acid	124049
Acetic acid, fluoro-, sodium salt	62748	Alanine, 3-[p-bis(2-chloroethyl)amino] phenyl-, L-	148823
Acetic acid, lead salt	301042	Aldicarb	116063
Acetic acid, thallium(I) salt	563688	Aldrin	309002
Acetic anhydride	108247	Allyl alcohol	107186
Acetimidic acid, N-[(methylcarbomoyl) oxy] thio. methyl ester	16752775	Allyl chloride	107051
		Aluminum phosphide	20859738
		Aluminum sulfate	10043013
Acetone	67641	5-(Aminomethyl)-3-isoxazolol	2763964
Acetone cyanohydrin	75865		

185

Hazardous Substance	CASRN
4-Aminopyridine	504245
Amitrole	61825
Ammonia	7664417
Ammonium acetate	631618
Ammonium benzoate	1863634
Ammonium bicarbonate	1066337
Ammonium bichromate	7789095
Ammonium bifluoride	1341497
Ammonium bisulfite	10192300
Ammonium carbamate	1111780
Ammonium carbonate	506876
Ammonium chloride	12125029
Ammonium chromate	7788989
Ammonium citrate, dibasic	3012655
Ammonium fluoborate	13826830
Ammonium fluoride	12125018
Ammonium hydroxide	1336216
Ammonium oxalate	6009707
	5972736
	14258492
Ammonium picrate	131748
Ammonium silicofluoride	16919190
Ammonium sulfamate	7773060
Ammonium sulfide	12135761
Ammonium sulfite	10196040
Ammonium tartrate	14307438
	3164292
Ammonium thiocyanate	1762954
Ammonium thiosulfate	7783188
Ammonium vanadate	7803556
Amyl acetate	628637
iso-	123922
sec-	626380
tert-	625161
Aniline	62533
Anthracene	120127
Antimony[b]	7440360
ANTIMONY AND COMPOUNDS	
Antimony pentachloride	7647189
Antimony potassium tartrte	28300745
Antimony tribromide	7789619
Antimony trichloride	10025919
Antimony trifluoride	7783564
Antimony trioxide	1309644

Hazardous Substance	CASRN
Aroclor 1016	12674112
Aroclor 1221	11104282
Aroclor 1232	11141165
Aroclor 1242	53469219
Aroclor 1248	12672296
Aroclor 1254	11097691
Aroclor 1260	11096825
Arsenic[b]	7440382
Arsenic acid	1327522
	7778394
ARSENIC AND COMPOUNDS	
Arsenic disulfide	1303328
Arsenic(III) oxide	1327533
Arsenic(V) oxide	1303282
Arsenic pentoxide	1303282
Arsenic trichloride	7784341
Arsenic trioxide	1327533
Arsenic trisulfide	1303339
Arsine, diethyl-	692422
Asbestos[c]	1332214
Auramine	492808
Azaserine	115026
Aziridine	151564
Azirino (2′,3″3,4)pyrrolo (1,2-a)indole-4, 7-dione, 6-amino-8-[((amino-carbonyl)oxy)methyl]-1,1a, 2,8,8a,8b-hexa-hydro-8a-methoxy-5-methyl	50077
Barium cyanide	542621
Benz(j)aceanthrylene, 1,2-dihydro-3-methyl-	56495
Benz(c)acridine	225514
3,4-Benzacridine	225514
Benzal chloride	98873
Benz(a)anthracene	56553
1,2-Benzanthracene	56553
1,2-Benzanthracene, 7,12-dimethyl-	57976
Benzenamine	62533
Benzenamine, 4,4′-carbonimidoylbis(N,N-dimethyl-	492808
Benzenamine, 4-chloro-	106478

Hazardous Substance	CASRN
Benzenamine, 4-chloro-2-methyl-, hydrochloride	3165933
Benzenamine, N,N-dimethyl-4-phenylazo-	60117
Benzenamine, 4,4'-methylenebis(2-chloro)-	101144
Benzenamine, 2-methyl-, hydrochloride	636215
Benzenamine, 2-methyl-5-nitro-	99558
Benzenamine, 4-nitro-	100016
Benzene	71432
Benzene, 1-bromo-4-phenoxy-	101553
Benzene chloro-	108907
Benzene, chloromethyl-	100447
Benzene, 1,2-dichloro-	95501
Benzene, 1,3-dichloro-	541731
Benzene, 1,4-dichloro-	106467
Benzene, dichloromethyl-	98873
Benzene, 2,4-diisocyanatomethyl	584849
	91087
	26471625
Benzene, dimethyl	1330207
m-	108383
o-	95476
p-	106423
Benzene, hexachloro-	118741
Benzene, hexahydro-	110827
Benzene, hydroxy-	108952
Benzene, methyl-	108883
Benzene, 1-methyl-2,4-dinitro-	121142
Benzene, 1-methyl-2,6-dinitro-	506202
Benzene, 1,2-methylenedioxy-4-allyl-	94597
Benzene, 1,2-methylenedioxy-4-propenyl-	120581
Benzene, 1,2 methylenedioxy-4-propyl	94586
Benzene, 1-methylethyl-	98828
Benzene, nitro-	98953

Hazardous Substance	CASRN
Benzene, pentachloro-	608935
Benzene, pentachloronitro-	82688
Benzene, 1,2,4,5-tetrachloro-	95943
Benzene, trichloromethyl-	98077
Benzene, 1,3,5-trinitro	99354
Benzeneacetic acid, 4-chloro-alpha-(4-chlorophenyl)-alpha-hydroxy, ethyl ester	510156
1,2-Benzenedicarboxylic acid anhydride	85449
1,2-Benzenedicarboxylic acid,[bis(2-ethylexyl)] ester	1178617
1,2-Benzenedicarboxylic acid, dibutyl ester	84742
1,2-Benzenedicarboxylic acid, diethyl ester	84662
1,2-Benzenedicarboxylic acid, dimethyl ester	131113
1,2-Benzenedicarboxylic acid, di-n-octyl ester	117840
1,3 Benzenediol	108463
1,2-Benzenediol,4-[1-hydroxy-2-(methylamino)ethyl]	51434
Benzenesulfonic acid chloride	98099
Benzenesulfonyl chloride	98099
Benzenethiol	108985
Benzidine	92875
1,2-Benzisothiazolin-3-one, 1,1-dioxide, and salts	81072
Benzo(a)anthracene	56553
Benzo(b)fluoranthene	205992
Benzo(k)fluoranthene	207089
Benzo(j,k)fluorene	206440
Benzoic acid	65850
Benzonitrile	100470
Benzo(ghi)perylene	191242
Benzo(a)pyrene	50328
3,4-Benzopyrene	50328
p-Benzoquinone	106514
Benzotrichloride	98077

Hazardous Substance	CASRN	Hazardous Substance	CASRN
Benzoyl chloride	98884	2-Butanone	78933
1,2-Benzphenanthrene	218019	2-Butanone peroxide	1338234
Benzyl chloride	100447	2-Butenal	123739
Beryllium[b]	7440417		4170303
BERYLLIUM AND COMPOUNDS		2-Butene, 1,4-dichloro-	764410
		Butyl acetate	123864
Beryllium chloride	7787475	iso-	110190
Beryllium dust	7440417	sec-	105464
Beryllium fluoride	7787497	tert-	540885
Beryllium nitrate	13597994	n-Butyl alcohol	71363
	7787555	Butylamine	109739
alpha - BHC	319846	iso-	78819
beta - BHC	319857	sec-	513495
gamma - BHC	58899	sec-	13952846
delta - BHC	319868	tert-	75649
2,2'-Bioxirane	1464535	Butyl benzyl phthalate	85687
(1,1'-Biphenyl)-4,4' diamine	92875	n-Butyl phthalate	84742
		Butyric acid	107926
(1,1'-Biphenyl)-4,4'-diamine,3,3' dichloro-	91941	iso-	79312
		Cacodylic acid	75605
(1,1'-Biphenyl)-4,4' diamine,3,3'dimethoxy-	119904	Cadmium	7440439
		Cadmium acetate	543908
(1,1' Biphenyl)-4,4'-diamine,3,3'-dimethyl-	119937	CADMIUM AND COMPOUNDS	
Bis(2-chloroethoxy) methane	111911	Cadmium bromide	7789426
		Cadmium chloride	10108642
Bis(2-chloroethyl) ether	111444	Calcium arsenate	7778441
Bis(2-chloroisopropyl) ether	108601	Calcium arsenite	52740166
		Calcium carbide	75207
Bis(chloromethyl) ether	542881	Calcium chromate	13765190
Bis(dimethylthiocarbamoyl) disulfide	137268	Calcium cyanide	592018
		Calcium dodecylbenzene sulfonate	26264062
Bis(2-ethylhexyl)phthalate	117817		
Bromine cyanide	506683	Calcium hypochlorite	7778543
Bromoacetone	598312	Camphene, octachloro-	8001352
Bromoform	75252	Captan	133062
4-Bromophenyl phenyl ether	101553	Carbamic acid, ethyl ester	51796
Brucine	357573	Carbamic acid, methylnitroso-, ethyl ester	615532
1,3-Butadiene, 1,1,2,3,4,4-hexachloro-	87683		
1-Butanamine, N-butyl-N-nitroso	924163	Carbamide, N-ethyl-N-nitroso-	759739
Butanoic acid, 4-[bis(2-chloroethyl)amino]benzene	305033	Carbamide, N-methyl-N-nitroso-	684935
1-Butanol	71363	Carbamide, thio-	62566

APPENDIX A

Hazardous Substance	CASRN
Carbamidoselenoic acid ...	630104
Carbamoyl chloride, dimethyl-	79447
Carbaryl	63252
Carbofuran	1563662
Carbon bisulfide	75150
Carbon disulfide	75150
Carbonic acid, dithallium (I) salt	6533739
Carbonochloridic acid, methyl ester	79221
Carbon oxyfluoride	353504
Carbon tetrachloride	56235
Carbonyl chloride	75445
Carbonyl fluoride	353504
Chloral	75876
Chlorambucil	305033
CHLORDANE (TECHNICAL MIXTURE AND METABOLITES)	
Chlordane	57749
Chlordane, technical	57749
CHLORINATED BENZENES	
CHLORINATED ETHANES	
CHLORINATED NAPHTHALENE	
CHLORINATED PHENOLS	
Chlorine	7782505
Chlorine cyanide	506774
Chlornaphazine	494031
Chloroacetaldehyde	107200
CHLOROALKYL ETHERS	
p-Chloroaniline	106478
Chlorobenzene	108907
4-Chloro-m-cresol	50507
p-Chloro-m-cresol	59507
Chlorodibromomethane ...	124481
1-Chloro-2,3-epoxypropane	106898
Chloroethane	75003
2-Chloroethyl vinyl ether ..	110758
Chloroform	67663

Hazardous Substance	CASRN
Chloromethyl methyl ether	107302
beta-Chloronaphthalene ...	91587
2-Chloronaphthalene	91587
2-Chlorophenol	95578
o-Chlorophenol	95578
4-Chlorophenyl phenyl ether	70057232
1-(o-Chlorophenyl)-thiourea	5344821
3-Chloropropionitrile	542787
Chlorosulfonic acid	7790945
4-Chloro-o-toluidine, hydrochloride	3165933
Chlorpyrifos	2921882
Chromic acetate	1066304
Chromic acid	11115745 7738945
Chromic acid, calcium salt	13765190
Chromic sulfate	10101538
Chromium^b	7440473
CHROMIUM AND COMPOUNDS	
Chromous chloride	10049055
Chrysene	218019
Cobaltous bromide	7789437
Cobaltous formate	544183
Cobaltous sulfamate	14017415
Coke Oven Emissions ...	N.A.
Copper^b	7440508
COPPER AND COMPOUNDS	
Copper cyanide	544923
Coumaphos	56724
Creosote	8001589
Cresol(s)	1319773
m-	108394
o-	95487
p-	106445
Cresylic acid	1319773
m-	108394
o-	95487
p-	106445
Crotonaldehyde	123739

Hazardous Substance	CASRN
	4170303
Cumene	98828
Cupric acetate	142712
Cupric acetoarsenite	12002038
Cupric chloride	7447394
Cupric nitrate	3251238
Cupric oxalate	5893663
Cupric sulfate	7758987
Cupric sulfate ammoniated	10380297
Cupric tartrate	815827
CYANIDES	
Cyanides (soluble cyanide salts), not elsewhere specified	57125
Cyanogen	460195
Cyanogen bromide	506683
Cyanogen chloride	506774
1,4-Cyclohexadienedione	106514
Cyclohexane	110827
Cyclohexanone	108941
1,3-Cyclopentadiene, 1,2,3,4,5,5-hexachloro-	77474
Cyclophosphamide	50180
2,4-D Acid	94757
2,4-D Esters	94111
	94791
	94804
	1320189
	1928387
	1928616
	1929733
	2971382
	25168267
	53467111
2,4-D, salts and esters	94757
Daunomycin	20830813
DDD	72548
4,4′ DDD	72548
DDE	72559
4,4′ DDE	72559
DDT	50293
4,4′ DDT	50293
DDT AND METABOLITES	
Decachloroctahydro-1,3,4-metheno-2H-	

Hazardous Substance	CASRN
cyclobuta[c,d]-pentalen-2-one	143500
Diallate	2303164
Diamine	302012
Diaminotoluene	95807
	25376458
	496720
	823405
Diazinon	5333415
Dibenz[a,h]anthracene	53703
1,2:5,6-Dibenzanthracene	53703
Dibenzo[a,h]anthracene	53703
1,2:7,8-Dibenzopyrene	189559
Dibenz[a,i]pyrene	189559
1,2-Dibromo-3-chloropropane	96128
Dibutyl phthalate	84742
Di-n-butyl phthalate	84742
Dicamba	1918009
Dichlobenil	1194656
Dichlone	117806
S-(2,3-Dichloroallyl) diisopropylthiocarbamate	2303164
3,5-Dichloro-N-(1,1-dimethyl-2-propynyl)benzamide	23950585
Dichlorobenzene (mixed)	25321226
1,2-Dichlorobenzene	95501
1,3-Dichlorobenzene	541731
1,4-Dichlorobenzene	106467
m-Dichlorobenzene	541731
o-Dichlorobenzene	95501
p-Dichlorobenzene	106467
DICHLOROBENZIDINE	
3,3′-Dichlorobenzidine	91941
Dichlorobromomethane	75274
1,4-Dichloro-2-butene	764410
Dichlorodifluoromethane	75718
Dichlorodiphenyl dichloroethane	72548
Dichlorodiphenyl trichloroethane	50293
1,1-Dichloroethane	75343
1,2-Dichloroethane	107062
1,1-Dichloroethylene	75354

APPENDIX A 191

Hazardous Substance	CASRN
1,2-trans-Dichloroethylene	156605
Dichloroethyl ether	111444
2,4-Dichlorophenol	120832
2,6-Dichlorophenol	87650
2,4-Dichlorophenoxyacetic acid, salts and esters	94757
Dichlorophenylarsine	606286
Dichloropropane	26638197
1,1-Dichloropropane	78999
1,3-Dichloropropane	142289
1,2-Dichloropropane	78875
Dichloropropane - Dichloropropene (mixture)	8003198
Dichloropropene	26952238
2,3-Dichloropropene	78886
1,3-Dichloropropene	542756
2,2-Dichloropropionic acid	75990
Dichlorvos	62737
Dieldrin	60571
1,2:3,4-Diepoxybutane	1464535
Diethylamine	109897
Diethylarsine	692422
1,4-Diethylene dioxide	123911
N,N'-Diethylhydrazine	1615801
O,O-Diethyl S-[2-(ethylthio)ethyl]phosphorodithioate	298044
O,O-Diethyl S-methyl dithiophosphate	3288582
Diethyl-p-nitrophenyl phosphate	311455
Diethyl phthalate	84662
O,O-Diethyl O-pyrazinyl phosphorothioate	297972
Diethylstilbestrol	56531
1,2-Dihydro-3,6-pyridazinedione	123331
Dihydrosafrole	94586
Diisopropyl fluorophosphate	55914
Dimethoate	60515
3,3'-Dimethoxybenzidine	119904
Dimethylamine	124403
Dimethylaminoazobenzene	60117

Hazardous Substance	CASRN
7,12-Dimethylbenz(a)anthracene	57976
3,3'-Dimethylbenzidine	119937
alpha,alpha-Dimethylbenzylhydroperoxide	80159
3,3-Dimethyl-1-(methylthio)-2-butanone,O-[(methylamino)carbonyl]oxime	39196184
Dimethylcarbamoyl chloride	79447
1,1-Dimethylhydrazine	57147
1,2-Dimethylhydrazine	540738
O,O-Dimethyl O-p-nitrophenyl phosphorothioate	298000
Dimethylnitrosamine	62759
alpha,alpha-Dimethylphenethylamine	122098
2,4-Dimethylphenol	105679
Dimethyl phthalate	131113
Dimethyl sulfate	77781
Dinitrobenzene (mixed)	25154545
m-	99650
o-	528290
p-	100254
4,6-Dinitro-o-cresol and salts	534521
4,6-Dinitro-o-cyclohexylphenol	131895
Dinitrophenol	25550587
2,5-	329715
2,6-	573568
2,4-Dinitrophenol	51285
Dinitrotoluene	25321146
3,4-Dinitrotoluene	610399
2,4-Dinitrotoluene	121142
Dinoseb	88857
Di-n-octyl phthalate	117840
1,4-Dioxane	123911
DIPHENYLHYDRAZINE	
1,2-Diphenylhydrazine	122667
Diphosphoramide, octamethyl-	152169
Dipropylamine	142847
Di-n-propylnitrosamine	621647

Hazardous Substance	CASRN
Diquat	85007
	2764729
Disulfoton	298044
2,4-Dithiobiuret	541537
Dithiopyrophosphoric acid, tetraethyl ester	3689245
Diuron	330541
Dodecylbenzenesulfonic acid	27176870
Endosulfan	115297
alpha-Endosulfan	959988
beta-Endosulfan	33213659
ENDOSULFAN AND METABOLITES	
Endosulfan sulfate	1031078
Endothall	145733
Endrin	72208
Endrin aldehyde	7421934
ENDRIN AND METABOLITES	
Epichlorohydrin	106898
Epinephrine	51434
Ethanal	75070
Ethanamine, 1,1-dimethy-2-phenyl-	122098
Ethanamine, N-ethyl-N-nitroso-	55185
Ethane, 1,2-dibromo-	106934
Ethane, 1,1-dichloro	75343
Ethane, 1,2-dichloro-	107062
Ethane, 1,1,1,2,2,2-hexachloro-	67721
Ethane, 1,1'-[methylenebis(oxy)]bis(2-chloro-	111911
Ethane, 1,1'-oxybis-	60297
Ethane, 1,1'-oxybis(2-chloro-	111444
Ethane, pentachloro-	76017
Ethane, 1,1,1,2-tetrachloro-	630206
Ethane, 1,1,2,2-tetrachloro-	79345
Ethane, 1,1,2-trichloro-	79005
Ethane, 1,1,1-trichloro-	

Hazardous Substance	CASRN
2,2-bis(p-methoxyphenyl)-	72435
1,2-Ethanediylbis-carbamodithioic acid	111546
Ethanenitrile	75058
Ethanethioamide	62555
Ethanol, 2,2'-(nitrosoimino)bis-	1116547
Ethanone, 1-phenyl-	98862
Ethanoyl chloride	75365
Ethenamine, N-methyl-N-nitroso-	4549400
Ethene, chloro-	75014
Ethene, 2-chloroethoxy-	110758
Ethene, 1,1-dichloro-	75354
Ethene, 1,1,2,2-tetrachloro-	127184
Ethene, trans- 1,2-dichloro-	156605
Ethion	583122
Ethyl acetate	141786
Ethyl acrylate	140885
Ethylbenzene	100414
Ethyl carbamate (Urethan)	51796
Ethyl cyanide	107120
Ethyl 4,4'-dichlorbenzilate	510156
Ethylene dibromide	106934
Ethylene dichloride	107062
Ethylene oxide	75218
Ethylenebis(dithiocarbamic acid)	111546
Ethylenediamine	107153
Ethylenediamine tetraacetic acid (EDTA)	60004
Ethylenethiourea	96457
Ethylenimine	151564
Ethyl ether	60297
Ethylidene dichloride	75343
Ethyl methacrylate	97632
Ethyl methanesulfonate	62500
Famphur	52857
Ferric ammonium citrate	1185575
Ferric ammonium oxalate	2944674

APPENDIX A 193

Hazardous Substance	CASRN
	55488874
Ferric chloride	7705080
Ferric dextran	9004664
Ferric fluoride	7783508
Ferric nitrate	10421484
Ferric sulfate	10028225
Ferrous ammonium sulfate	10045893
Ferrous chloride	7758943
Ferrous sulfate	7720787
	7782630
Fluoroacetic acid, sodium salt	62748
Fluoranthene	206440
Fluorene	86737
Fluorine	7782414
Fluoroacetamide	640197
Formaldehyde	50000
Formic acid	64186
Fulminic acid, mercury (II) salt	628864
Fumaric acid	110178
Furan	110009
Furan, tetrahydro-	109999
2-Furancarboxaldehyde	98011
2,5-Furandione	108316
Furfural	98011
Furfuran	110009
D-Glucopyranose, 2-deoxy-2-(3-methyl-3-nitrosoureido)-	18883664
Glycidylaldehyde	765344
Guanidine, N-nitroso-N-methyl-N'-nitro-	70257
Guthion	86500
HALOETHERS	
HALOMETHANES	
Heptachlor	76448
HEPTACHLOR AND METABOLITES	
Heptachlor epoxide	1024573
Hexachlorobenzene	118741
Hexachlorobutadiene	87683
HEXACHLOROCY-CLOHEXANE (all isomers)	608731

Hazardous Substance	CASRN
Hexachlorocyclohexane (gamma isomer)	58899
Hexachlorocyclopentadiene	77474
1,2,3,4,10,10-Hexachloro-6,7-epoxy-1,4,4a,5,6,7,8,8a-octahydro-endo-endo-1,4:5,8-dimethanonaphthalene	72208
1,2,3,4,10,10-Hexachloro-6,7-epoxy-1,4,4a,5,6,7,8,8a-octahydro-endo-exo-1,4:5,8-dimethanonaphthalene	60571
Hexachloroethane	67721
Hexachlorohexahydro-endo,endo-dimethanonaphthalene	465736
1,2,3,4,10,10-Hexachloro-1,4,4a,5,8,8a-hexahydro-1,4,5,8-endo,endo-dimethanonaphthalene	465736
1,2,3,4,10,10-Hexachloro-1,4,4a,5,8,8a-hexahydro-1,4:5,8-endo,exo-dimethanonaphthalene	309002
Hexachlorophene	70304
Hexachloropropene	1888717
Hexaethyl tetraphosphate	757584
Hydrazine	302012
Hydrazine, 1,2-diethyl-	1615801
Hydrazine, 1,1-dimethyl-	57147
Hydrazine, 1,2-dimethyl-	540738
Hydrazine, 1,2-diphenyl-	122667
Hydrazine, methyl-	60344
Hydrazinecarbothioamide	79196
Hydrochloric acid	7647010
Hydrocyanic acid	74908
Hydrofluoric acid	7664393
Hydrogen cyanide	74908
Hydrogen fluoride	7664393
Hydrogen phosphide	7803512
Hydrogen sulfide	7783064

Hazardous Substance	CASRN	Hazardous Substance	CASRN
Hydroperoxide, 1-methyl-1-phenylethyl-	80159	Maleic acid	110167
Hydrosulfuric acid	7783064	Maleic anhydride	108316
Hydroxydimethylarsine oxide	75605	Maleic hydrazide	123331
2-Imidazolidinethione	96457	Malononitrile	109773
Indeno(1,2,3-cd)pyrene	193395	Melphalan	148823
Iron dextran	9004664	Mercaptodimethur	2032657
Isobutyl alcohol	78831	Mercuric cyanide	592041
Isocyanic acid, methyl ester	624839	Mercuric nitrate	10045940
Isophorone	78591	Mercuric sulfate	7783359
Isoprene	78795	Mercuric thiocyanate	592858
Isopropanolamine dodecylbenzenesulfonate	42504461	Mercurous nitrate	10415755 7782867
Isosafrole	120581	Mercury	7439976
3(2H)-Isoxazolone, 5-(aminomethyl)-	2763964	MERCURY AND COMPOUNDS	
Kelthane	115322	Mercury, (acetato-O)phenyl-	62384
Kepone	143500	Mercury fulminate	628864
Lasiocarpine	303344	Methacrylonitrile	126987
Lead[b]	7439921	Methanamine, N-methyl-	124403
Lead acetate	301042	Methane, bromo-	74839
LEAD AND COMPOUNDS		Methane, chloro-	74873
Lead arsenate	7784409 7645252 10102484	Methane, chloromethoxy-	107302
		Methane, dibromo-	74953
		Methane, dichloro-	75092
		Methane, dichlorodifluoro-	75718
Lead chloride	7758954	Methane, iodo-	74884
Lead fluoborate	13814965	Methane, oxybis(chloro)-	542881
Lead fluoride	7783462	Methane, tetrachloro-	56235
Lead iodide	10101630	Methane, tetranitro-	509148
Lead nitrate	10099748	Methane, tribromo-	75252
Lead phosphate	7446277	Methane, trichloro-	67663
Lead stearate	7428480 1072351 56189094 52652592	Methane, trichlorofluoro-	75694
		Methanesulfonic acid, ethyl ester	62500
		Methanethiol	74931
		Methanesulfenyl chloride, trichloro-	594423
Lead subacetate	1335326		
Lead sulfate	15739807 7446142	4,7-Methano-1H-indene, 1,4,5,6,7,8,8-heptachloro-3a,3,7,7a-tetrahydro-	76448
Lead sulfide	1314870		
Lead thiocyanate	592870		
Lindane	58899	Methanoic acid	64186
Lithium chromate	14307358	4,7-Methanoindan, 1,2,4,5,6,7,8,8-	
Malathion	121755		

APPENDIX A

Hazardous Substance	CASRN
octachloro-3a,4,7,7a-tetrahydro-	57749
Methanol	67561
Methapyrilene	91805
Methomyl	16752775
Methoxychlor	72435
Methyl alcohol	67561
2-Methylaziridine	75558
Methyl bromide	74839
1-Methylbutadiene	504609
Methyl chloride	74873
Methyl chlrorocarbonate	79221
Methyl chloroform	71556
4,4'-Methylenebis(2-chloroaniline)	101144
2,2'-Methylenebis(3,4,6-trichlorophenol)	70304
3-Methylcholanthrene	56495
Methylene bromide	74953
Methylene chloride	75092
Methylene oxide	50000
Methyl ethyl ketone	78933
Methyl ethyl ketone peroxide	1338234
Methyl hydrazine	60344
Methyl iodide	74884
Methyl isobutyl ketone	108101
Methyl isocyanate	624839
2-Methyllactonitrile	75865
Methylmercaptan	74931
Methyl methacrylate	80626
N-Methyl-N'-nitro-N-nitrosoguanidine	70257
Methyl parathion	298000
4-Methyl-2-pentanone	108101
Methylthiouracil	56042
Mevinphos	7786347
Mexacarbate	315184
Mitomycin C	50077
Monoethylamine	75047
Monomethylamine	74895
Naled	300765
5,12-Naphthacenedione, (8S-cis)-8-acetyl-10-[3-amino-2,3,6-trideoxy-alpha-L-lyxo-hexopyranosyl)oxy]-7,8,9,10-tetrahydro-6,8,11-trihydroxy-1-methoxy-	20830813
Naphthalene, 2-chloro	91203
Naphthalene, 2-chloro-	91587
1,4-Naphthalenedione	130154
2,7-Naphthalenedisulfonic acid, 3,3'-[(3,3'-dimethyl-(1,1'-biphenyl)-4,4'-diyl)-bis(azo)]bis(5-amino-4-hydroxy)-tetrasodium salt	72571
Naphthenic acid	1338245
1,4-Naphthoquinone	130154
1-Naphthylamine	134327
2-Naphthylamine	91598
alpha-Naphthylamine	134327
beta-Naphthylamine	91598
2-Naphthylamine, N,N-bis(2-chloroethyl)-	494031
alpha-Naphthylthathiourea	86884
Nickel[b]	7440020
NICKEL AND COMPOUNDS	
Nickel ammonium sulfate	16599180
Nickel carbonyl	13463393
Nickel chloride	7718549
	37211055
Nickel cyanide	557197
Nickel(II) cyanide	557197
Nickel hydroxide	12054487
Nickel nitrate	14216752
Nickel sulfate	7786814
Nickel tetracarbonyl	13463393
Nicotine and salts	54115
Nitric acid	7697372
Nitric oxide	10102439
p-Nitroaniline	100016
Nitrobenzene	98953
Nitrogen dioxide	10102440

Hazardous Substance	CASRN	Hazardous Substance	CASRN
	10544726	1,2-Oxathiolane, 2,2-	
Nitrogen(II) oxide	10102439	dioxide	1120714
Nitrogen(IV) oxide	10102440	2H-1,3,2-	
	10544726	Oxazaphosphorine, 2-	
Nitroglycerine	55630	[bis(2-chloroethyl)	
Nitrophenol (mixed)	25154556	amino]tetrahydro-2-	
m-	554847	oxide	50180
o-	88755	Oxirane	75218
p-	100027	Oxirane, 2-	
p-Nitrophenol	100027	(chloromethyl)-	106898
2-Nitrophenol	88755	Paraformaldehyde	30525894
4-Nitrophenol	100027	Paraldehyde	123637
NITROPHENOLS		Parathion	56382
2-Nitropropane	79469	Pentachlorobenzene	608935
NITROSAMINES		Pentachloroethane	76017
N-Nitrosodi-n-butylamine	924163	Pentachloronitrobenzene	82688
N-Nitrosodiethanolamine	1116547	Pentachlorophenol	87865
N-Nitrosodiethylamine	55185	1,3-Pentadiene	504609
N-Nitrosodimethylamine	62759	Phenacetin	62442
N-Nitrosodiphenylamine	86306	Phenanthrene	85018
N-Nitrosodi-n-		Phenol	108952
propylamine	621647	Phenol, 2-chloro-	95578
N-Nitroso-N-ethylurea	759739	Phenol, 4-chloro-3-	
N-Nitroso-N-methylurea	684935	methyl-	59507
N-Nitroso-N-		Phenol, 2-cyclohexyl-4,6-	
methylurethane	615532	dinitro-	131895
N-Nitrosometh-		Phenol, 2,4-dichloro-	120832
ylvinylamine	4549400	Phenol, 2,6-dichloro-	87650
N-Nitrosopiperidine	100754	Phenol, 2,4-dimethyl-	105679
N-Nitrosopyrrolidine	930552	Phenol, 2,4-dinitro-	51285
Nitrotoluene	1321126	Phenol, 2,4,-dinitro-6-(1-	
m-	99081	methylpropyl)-	88857
o-	88722	Phenol, 2,4-dinitro-6-	
p-	99990	methyl-, and salts	534521
5-Nitro-o-toluidine	99558	Phenol, 4-nitro-	100027
5-Norbornene-2,3-		Phenol, pentachloro-	87865
dimethanol, 1,4,5,6,7,7-		Phenol, 2,3,4,6-	
hexachloro, cyclic		tetrachloro-	58902
sulfite	115297	Phenol, 2,4,5-trichloro-	95954
Octamethylpyrophosphoramide		Phenol, 2,4,6-trichloro-	88062
	152169	Phenol, 2,4,6-trinitro-,	
Osmium oxide	20816120	ammonium salt	131748
Osmium tetroxide	20816120	Phenyl dichloroarsine	696286
7-		1,10-(1,2-	
Oxabicyclo[2,2,1]heptane-		Phenylene)pyrene	193395
2,3-dicarboxylic acid	145733	Phenylmercuric acetate	62384

APPENDIX A   197

Hazardous Substance	CASRN
N-Phenylthiourea	103855
Phorate	298022
Phosgene	75445
Phosphine	7803512
Phosphoric acid	7664382
Phosphoric acid, diethyl p-nitrophenyl ester	311455
Phosphoric acid, lead salt	7446277
Phosphorodithioic acid, O,O-diethyl S-methylester	3288582
Phosphorodithioic acid, O,O-diethyl S-(ethylthio), methyl ester	298022
Phosphorodithioic acid, O,O-dimethyl S-2(methylamino)-2-oxoethyl ester	60515
Phosphorofluoridic acid, bis(1-methylethyl) ester	55914
Phosphorothioic acid, O,O-diethyl O-(p-nitrophenyl) ester	56382
Phosphorothioic acid, O,O-diethyl O-pyrazinyl ester	297972
Phosphorothioic acid, O,O-dimethyl O-{p-[(dimethylamino)sulfonyl]phenyl} ester	52857
Phosphorus	7723140
Phosphorus oxychloride	10025873
Phosphorus pentasulfide	1314803
Phosphorus sulfide	1314803
Phosphorus trichloride	7719122
PHTHALATE ESTERS	
Phthalic anhydride	85449
2-Picoline	109068
Plumbane, tetraethyl-	78002
POLYCHLORINATED BIPHENYLS (PCBs)	1336363
	12674112
	11104282
	11141165
	53469219
	12672296

Hazardous Substance	CASRN
	11097691
	11096825
POLYNUCLEAR AROMATIC HYDROCARBONS	
Potassium arsenate	7784410
Potassium arsenite	10124502
Potassium bichromate	7778509
Potassium chromate	7789006
Potassium cyanide	151508
Potassium hydroxide	1310583
Potassium permanganate	7722647
Potassium silver cyanide	506616
Pronamide	23950585
1-Propanal, 2,3-epoxy-	765344
Propanal, 2-methyl-2-(methylthio)-,O-[(methylamino) carbonyl]oxime	116063
1-Propanamine	107108
1-Propanamine, N-propyl-	142847
Propane, 1,2-dibromo-3-chloro-	96128
Propane, 2-nitro-	79469
Propane, 2,2′-oxybis(2-chloro-	108601
1,3-Propane sulfone	1120714
Propanedinitrile	109773
Propanenitrile	107120
Propanenitrile, 3-chloro-	542767
Propanenitrile, 2-hydroxy-2-methyl-	75865
1,2,3-Propanetriol, trinitrate-	55630
1-Propanol, 2,3-dibromo, phosphate (3:1)	126727
1-Propanol, 2-methyl-	78831
2-Propanone	67641
2-Propanone, 1-bromo	598312
Propargite	2312358
Propargyl alcohol	107197
2-Propenal	107028
Propenamide	79061
Propene, 1,3-dichloro-	542756

Hazardous Substance	CASRN
1-Propene, 1,1,2,3,3,3-hexachloro-	1888717
2-Propenenitrile	107131
2-Propenenitrile, 2-methyl-	126987
2-Propenoic acid	79107
2-Propenoic acid, ethyl ester	140885
2-Propenoic acid, 2-methyl-, ethyl ester	97632
2-Propenoic acid, 2-methyl-, methyl ester	80626
2-Propen-1-ol	107186
Propionic acid	79094
Propionic acid, 2-(2,4,5-trichlorophenoxy)-	93721
Propionic anhydride	123626
n-Propylamine	107108
Propylene dichloride	78875
Propylene oxide	75569
1,2-Propylenimine	75558
2-Propyn-1-ol	107197
Pyrene	129000
Pyrethrins	121299
	121211
	8003347
4-Pyridinamine	504245
Pyridine	110861
Pyridine, 2-[(2-(dimethylamino)ethyl)-2-thenylamino]-	91805
Pyridine, hexahydro-N-nitroso-	100754
Pyridine, 2-methyl-	109068
Pyridine, (S)-3-(1-methyl-2-pyrrolidinyl)-, and salts	54115
4(1H)-Pyrimidinone, 2,3-dihydro-6-methyl-2-thioxo-	56042
Pyrophosphoric acid, tetraethyl ester	107493
Pyrrole, tetrahydro-N-nitroso-	930552
Quinoline	91225
RADIONUCLIDES	
Reserpine	50555
Resorcinol	108463
Saccharin and salts	81072
Safrole	94597
Selenious acid	7783008
Selenium[b]	7782492
SELENIUM AND COMPOUNDS	
Selenium dioxide	7446084
Selenium disulfide	7488564
Selenium oxide	7446084
Selenourea	630104
L-Serine, diazoacetate (ester)	115026
Silver[b]	7440224
SILVER AND COMPOUNDS	
Silver cyanide	506649
Silver nitrate	7761888
Silvex	93721
Sodium	7440235
Sodium arsenate	7631892
Sodium arsenite	7784465
Sodium azide	26628228
Sodium bichromate	10588019
Sodium bifluoride	1333831
Sodium bisulfite	7831905
Sodium chromate	7775113
Sodium cyanide	143339
Sodium dodecylbenzene sulfonate	25155300
Sodium fluoride	7881494
Sodium hydrosulfide	16721805
Sodium hydroxide	1310732
Sodium hypochlorite	7681529
	10022705
Sodium methylate	124414
Sodium nitrite	7632000
Sodium phosphate, dibasic	7558794
	10039324
	10140655
Sodium phosphate, tribasic	7601549
	7785844
	10101890

Hazardous Substance	CASRN	Hazardous Substance	CASRN
	10361894	1,1,1,2-Tetrachloroethane	630206
	7758294	1,1,2,2-Tetrachloroethane	79345
	10124568	Tetrachloroethylene	127184
Sodium selenite	10102188	2,3,4,6-	
	7782823	Tetrachlorophenol	58902
4,4'-Stilbenediol, alpha,		Tetraethyldithio-	
alpha'-diethyl-	56531	pyrophosphate	3689245
Streptozotocin	18883664	Tetraethyl lead	78002
Strontium chromate	7789062	Tetraethyl pyrophosphate	107493
Strontium sulfide	1314961	Tetrahydrofuran	109999
Strychnidin-10-one, and		Tetranitromethane	509148
salts	57249	Tetraphosphoric acid,	
Strychnidin-10-one, 2,3-		hexaethyl ester	757584
dimethoxy-	357573	Thallic oxide	1314325
Strychnine and salts	57249	Thallium[b]	7440280
Styrene	100425	THALLIUM AND	
Sulfur hydride	7783064	COMPOUNDS	
Sulfur monochloride	12771083	Thallium(I) acetate	563688
Sulfur phosphide	1314803	Thallium(I) carbonate	6533739
Sulfur selenide	7488564	Thallium(I) chloride	7791120
Sulfuric acid	7664939	Thallium(I) nitrate	10102451
	8014957	Thallium(III) oxide	1314325
Sulfuric acid, dimethyl		Thallium(I) selenide	12039520
ester	77781	Thallium(I) sulfate	7446186
Sulfuric acid, thallium(I)			10031591
salt	7446186	Thioacetamide	62555
	10031591	Thiofanox	39196184
2,4,5,-T	93765	Thiomidodicarbonic	
2,4,5,-T acid	93765	diamide	541537
2,4,5-T amines	2008460	Thiomethanol	74931
	6369966	Thiophenol	108985
	6369977	Thiosemicarbazide	79196
	1319728	Thiourea	62566
	3813147	Thiourea,	
2,4,5-T esters	93798	(2,chlorophenyl)-	5344821
	2545597	Thiourea, 1-	
	61792072	naphthalenyl-	86884
	1928478	Thiourea, phenyl-	103855
	25168154	Thiram	137268
2,4,5-T salts	13560991	Toluene	108883
TDE	72548	Toluenediamine	95807
1,2,4,5-			25376458
Tetrachlorobenzene	95943		496720
2,3,7,8-			823405
Tetrachlorodibenzo-p-		Toluene diisocyanite	584849
dioxin (TCDD)	1746016		91087

Hazardous Substance	CASRN	Hazardous Substance	CASRN
o-Toluidine hydrochloride	26471625	Cadmium	
	636215	Chromium	
Toxaphene	8001352	Lead	
2,4,5-TP acid	93721	Mercury	
2,4,5-TP acid esters	32534955	Selenium	
1H-1,2,4-Triazol-3-amine	61825	Silver	
Trichlorfon	52686	Endrin	
1,2,4-Trichlorobenzene	120821	Lindane	
1,1,1-Trichloroethane	71556	Methoxychlor	
1,1,2-Trichloroethane	79005	Toxaphene	
Trichloroethene	79016	2,4-D	
Trichlorethylene	79016	2,4,5-TP	
Trichloromethanesulfenyl chloride	594423	Uracil 5-[bis(2-chloroethyl)amino]-	66751
Trichloromonofluoromethane	75694	Uracil mustard	66751
		Uranyl acetate	541093
Trichlorophenol	25167822	Uranyl nitrate	10102064
2,3,4-Trichlorophenol	15950660		36478769
2,3,5-Trichlorophenol	933788	Vanadic acid, ammonium salt	7803556
2,3,6-Trichlorophenol	933755	Vanadium(V) oxide	1314621
2,4,5-Trichlorophenol	95954	Vanadium pentoxide	1314621
2,4,6-Trichlorophenol	88062	Vanadyl sulfate	27774136
3,4,5-Trichlorophenol	609198	Vinyl acetate	108054
2,4,5-Trichlorophenol	95954	Vinyl chloride	75014
2,4,6-Trichlorophenol	88062	Vinylidene chloride	75354
2,4,5-Trichlorophenoxyacetic acid	93765	Warfarin	81812
		Xylene (mixed)	1330207
Triethanolamine dodecylbenzenesulfonate	27323417	m-	108383
Triethylamine	121448	o-	95476
Trimethylamine	75503	p-	106423
sym-Trinitrobenzene	99354	Xylenol	1300716
1,3,5-Trioxane, 2,4,6-trimethyl-	123637	Yohimban-16-carboxylic acid, 11,17-dimethoxy-18-[(3,4,5-trimethoxybenzoyl)oxy]-, methylester	50555
Tris(2,3-dibromopropyl) phosphate	126727	Zinc[b]	7440666
Trypan blue	72571	ZINC AND COMPOUNDS	
Unlisted Hazardous Wastes		Zinc acetate	557346
Characteristic of Ignitability		Zinc ammonium chloride	52628258
Characteristic of Corrosivity			14639975
Characteristic of Reactivity			14639986
Characteristic of EP Toxicity		Zinc borate	1332076
Arsenic		Zinc bromide	7699458
Barium		Zinc carbonate	3486359

Hazardous Substance	CASRN	Hazardous Substance	CASRN
Zinc chloride	7646857	Zinc silicofluoride	16871719
Zinc cyanide	557211	Zinc sulfate	7733020
Zinc fluoride	7783495	Zirconium nitrate	13746899
Zinc formate	557415	Zirconium potassium fluoride	16923958
Zinc hydrosulfite	7779864		
Zinc nitrate	7779886	Zirconium sulfate	14644612
Zinc phenolsulfonate	127822	Zirconium tetrachloride	10026116
Zinc phosphide	1314847		

[a] Chemical Abstracts Service Registry Number.

[b] No reporting of releases of the hazardous substance is required if diameter of the pieces of the solid metal released is equal to or exceeds 110 μm (0.004 in.).

[c] The reportable quantity for asbestos is limited to friable forms only.

APPENDIX B

Storage System Gauging Procedures

(Source: American Petroleum Institute Publication 1621, "Recommended Practices for Bulk Liquid Stock Control at Retail Outlets" [third edition, 1977]).

Insert the pole through the gauge hole of the tank until the tip touches the tank bottom. The pole should be inserted at the same point in the gauge hole each time a gauge is taken and should be held in a vertical position. Be sure that it does not rest on a projection on the tank bottom. Withdraw the pole quickly to avoid creepage of the product, and read the product "cut" on the graduated scale to the nearest $1/8$ inch. When gasoline or another volatile product is gauged, the reading adjacent to the "cut" on the grooved portion of the pole should be taken as the gauge. Clean the pole at the "cut" by wiping with a cloth and repeat the procedure to check the accuracy of the reading.

A water-finding paste, which is unaffected by gasoline, but which will change color in water, is used to check for water at the bottom of storage tanks. Information on satisfactory pastes may be obtained from the supplier.

It is used as follows:

> Coat the end of the gauge stick on the graduated side with a light, even film of the paste for approximately 3 in. Insert the pole through the gauge hole until the pole reaches the bottom of the tank. Be sure that the pole is not resting on an obstruction or other projection on the tank bottom. Keep the pole in this position for the time specified for the product. Then withdraw it and read the water "cut" (as noted by change in color of the paste) on the graduated scale to the nearest $1/8$ in.
>
> The immersion time for a water "cut" is approximately 10 sec for light products such as gasoline and kerosene, and 20 to 30 sec for heavier products. If the test shows more than $1/2$ in. of water, arrange-

ments should be made for its immediate removal and the supplier notified.

USE OF TANK CALIBRATION CHART

After gauging the tank, select the correct calibration chart and proceed as follows:

1. Read chart directly for all gauges which are to the exact inch (tolerance ± $1/16$ in.).
2. For gauges of $1/8$ in. over or under the exact inch, proceed as follows:
   a. Read chart for exact inch gauge on scale above and below actual stick-gauge reading—for example, if stick gauge reads 46.5 inches, read chart at 46 in. and 47 in.
   b. Subtract gallonage shown on scale at these two readings. For example, for a 1,000-gal tank (diameter 64 in., length 72 in.):

   $$\begin{aligned} \text{Chart reading at 47 in.} &= 789 \text{ gal} \\ \text{Chart reading at 46 in.} &= 771 \text{ gal} \\ \text{Subtracting} &= 18 \text{ gallons} \end{aligned}$$

   c. Multiply this gallonage by the fraction of an inch, e.g.:

   $$18 \text{ gal} \times 1/2 = 9 \text{ gallons}$$

   d. Add the gallonage shown on the chart for the lower whole-inch gauge and the gallons calculated by step (c), e.g.:

   $$\begin{aligned} \text{Gal at 46 in.} &= 771 \text{ gal} \\ \text{Gal at } 1/2 \text{ in.} &= 9 \text{ gal} \\ \text{Adding} &= 780 \text{ gal} \end{aligned}$$

Therefore, the tank gauge of $46 1/2$ in. represents 780 gal of product in the 1,000-gal tank (diameter 64 in.; length 72 in.).

APPENDIX C

Equipment and Procedure for Testing Accuracy of Gasoline-Dispensing Meters

(Source: American Petroleum Institute Publication 1621, "Recommended Practices for Bulk Liquid Stock Control at Retail Outlets" [third edition, 1977]).

EQUIPMENT

A proving can of 5-gal capacity is the only special equipment required. The upper portion of this can consists of a neck, approximately 4 in. in diameter, having a sight glass with an adjacent scale graduated in in.^{3} above and below a zero point. It indicates the number of in.^{3} delivered by the meter greater than or less than the amount indicated on the dispenser dial.

If local regulations require the station operator to check dispensing units periodically, the local bureau of weights and measures should be consulted concerning the size and type of proving can to be used. Otherwise, a suitable can may be purchased from a reliable automotive equipment supplier. Each can should be checked for accuracy periodically and/or as may be required by local regulations.

PROCEDURE

1. Wet the can by filling with product to its full capacity, and return the product to the storage tank.
2. Refill the can to its capacity, as indicated by the dispenser, with the nozzle fully open (maximum filling rate).
3. Read on the graduated scale the number of in.^{3} delivered greater or less than the quantity shown on the dispenser. Note the difference.
4. Return the product in the can to storage.
5. Refill the can as in Step 2, but with the nozzle partly closed so that flow is limited to approximately 5 gal/min.

6. Repeat Steps 3 and 4.
7. If the quantity delivered in Step 2 or Step 5 varies by more than 7 in.^{3} above or below the zero point, adjustment by a qualified pump mechanic should be arranged. The operator should not attempt to adjust the meter himself.
8. Note in the inventory record the changed meter readings caused by the delivery of the product used for the test, noting also whether the product was returned to storage or used for other purposes.
9. Keep records of calibrations to assist in reconciling inventory variations.

APPENDIX D

State Underground Storage Regulatory Programs—Release Detection and Monitoring Requirements in California, Delaware, and Florida

CALIFORNIA

The California Underground Storage Regulations (California Administrative Code, Title 23, Chapter 3, Subchapter 16) provide for a number of options to achieve release detection and monitoring.

Section 2641 states that all owners or operators who cannot implement visual monitoring for the entire underground storage tank during all periods of the year must employ one of the following options:

1. Underground storage tank testing
   - monthly frequency
   - according to NFPA 329 criteria
   - hydrostatic testing permitted for piping
   - reporting to local agency
2. Vapor or other vadose zone monitoring and groundwater monitoring with soil sampling
   - soil sampling with well installation
   - vadose zone monitoring – continuous or daily
   - groundwater monitoring – semiannually (groundwater should be less than 100 ft from ground surface)
3. Vadose zone monitoring, soil sampling, and underground storage tank testing
   - soil sampling with well installation
   - vadose zone monitoring – continuous or daily
   - tank testing – annually (groundwater should be more than 100 ft from ground surface)
4. Groundwater and soil testing
   - soil sampling with well installation
   - groundwater monitoring – monthly (groundwater should be less than 30 ft from ground surface and not have any beneficial use)

5. Inventory reconciliation, underground storage tank testing, and pipeline leak detectors
   - inventory reconciliation – daily
   - tank testing – annual
   - pipeline leak detectors

   (Storage system inputs and withdrawals must be metered; allowable inventory reconciliation variance is 0.15% of throughput plus an allowable measurement error based on tank size)

6. Inventory reconciliation, underground storage tank testing, pipeline leak detectors, vadose zone, or groundwater monitoring and soil testing
   - inventory reconciliation – daily
   - tank testing – annual
   - pipeline leak detectors
   - soil sampling with well installation
   - vadose zone or groundwater monitoring

   (Limited to motor fuel storage systems; inventory reconciliation variations based upon number of measurements)

7. Underground storage tank gauging and testing
   - tank gauging – weekly
   - tank testing – annual

   (Limited to small tanks that do not have frequent inputs or withdrawals and where the liquid level can be measured to an accuracy of ± 5 gal)

8. Interim monitoring – underground storage tank testing and inventory reconciliation or tank gauging
   - tank testing – annual
   - inventory reconciliation – daily
   - tank gauging (small tanks)

   (Interim monitoring for only 3 years; small businesses and governmental agencies revert to Options 1-7; all others replace or close tankage)

DELAWARE

The Delaware Underground Storage Regulations (Delaware Code, Title 7, Chapter 74; Delaware Storage Tank Act, Regulations Governing Underground Storage Tanks) provide for several options that specifically address release detection and monitoring in new and existing storage systems.

New Storage Systems

All new storage systems must be provided with a means of monitoring for any release of the stored regulated substance and must include:

- inventory monitoring
- line leak detection system for positive pressure pumping systems

and one or more of the following:

- interstitial monitoring for double-walled tanks
- monitoring well or detector in an impervious secondary containment system
- continuous in-tank gauging system
- observation well system (a minimum of 4 wells installed within or outside the tank excavation)
- vadose zone monitoring
- U-tube monitoring system
- tank tightness testing according to the following schedule:

Tank Age	Testing Frequency
> 10 yrs	1 test/5 yr
> 15 yrs	1 test/3 yr
> 21 yrs	1 test/yr

Existing Storage Systems

These systems must provide:

- inventory monitoring
- line leak detection system for positive pressure pumping systems

and one of the following:

- tank tightness test as per the schedule listed above
- observation wells (a minimum of 3 wells)
- continuous in-tank gauging system

FLORIDA

The Florida Underground Storage Regulations (375.30 Florida Statutes, Chapter 17–61, "Stationary Tanks") also provide several options that specifically address release detection and monitoring in new and existing storage systems.

New Storage Systems

All new storage systems must be provided with a means of monitoring for any leakage and stored pollutant at the time of installation. The monitoring system must consist of

- inventory monitoring

and one or more of the following:

- continuous interstitial monitor
- monitoring well or detector in an impervious secondary containment
- continuous leak detection system in the tank excavation or around a tank
- monitoring wells (a minimum of 4 wells)
- groundwater monitoring plan
- SPCC plan

Existing Storage Systems

Existing storage systems must provide inventory monitoring. All existing systems must be retrofitted in accordance with section 17-61.06(2)(b), according to the schedule in Table 1.

Table 1. Schedule for Retrofitting of Existing Underground Storage Tanks, Florida Regulations[a]

Year Tank Installed	Year Retrofitting Required						
	1986	1987	1988	1989	1993	1995	1998
Prior to 1970	MO			LR			
1970–1975		MO			LR		
1976–1980			MO			LR	
1981–September 1, 1984				MO			LR

[a]MO = Installation of monitoring system and devices and overfill protection
LR = Lining or replacement of non-approved-type tanks

Index

accumulator systems 86
advection 59
American Petroleum Institute (API) 22, 32, 38, 40, 41, 135-136
American Society of Mechanical Engineers (ASME) 5
American Society for Testing and Materials (ASTM) 5
anode 5, 7-8, 12
automatic gauging 30, 46-48, 49-50

bailers 97-98
Barcol hardness 9
barometric pressure 141, 142
buoyancy 130-131
butt welding 5

cable detection system 173, 174
calibration chart 204
California regulations 207-208
catalytic detectors 106
cathodic protection 4
chemical/biological transformation 60
class I liquid 23, 24
coating 7, 11
coefficient of expansion 25, 37, 121, 123, 135-136
combustible gas indicator 106
Comprehensive Environmental Response, Compensation, and Liability Act (CERCLA) 16, 185-201
concrete encasement 181
condensation 147
continuous monitors 120

convection 63
copper-copper sulfate reference half cell 5
corrosion 5
corrosion resistant material 8

Delaware regulations 208-209
Department of Defense (DOD) 4
detection wells 65-73
detection tubes 105
differential float device 99
diffusion 63
direct testing 172
dispensing meter calibration 205-206
dispersion 60
double walled piping 177, 180-181
double walled tanks 10, 13
drilling methods 66-68
dual tracer 89
dyes and tracers 73, 78, 81

electrical resistivity 81-82
electrical resistivity sensor 100, 101
electronic pressure monitor 169-170
Environmental Protection Agency (EPA) 36, 38, 146, 160
  action numbers 36, 38
evaporation 28-29, 141, 147
external sensing 173

fiberglass reinforced plastic 9, 13, 144, 145
flame ionization detector 107
Florida regulations 209-210

fluctuation water table 62
fluorescein 78
flux chamber
  downhole 84, 86
  surface 84
Freon 78

galvanized steel 5
galvanic corrosion 5
gauge stick 30, 32, 42
gauging accuracy 30–31, 203–204
gas chromatography 152
geophysical techniques 81–83
ground probe testing 88–89
groundwater monitoring 173
groundwater sampling 74–80

head pressure 122, 125, 126, 128, 130, 136, 139, 144, 146, 147–149
helium 150–151
histogram 27, 28, 155
hydraulic piping testing 172
hydrocarbon permeable device 102–103
hydrocarbon sensitive pastes 97
hydrocarbon soluble device 101–102
hysteresis 62

immiscible liquids 59
impervious barrier 176–179
impressed current 5, 7, 12
indirect testing 172–173
interface probe 99
interferometer 128, 129, 132
integrity testing (tank shell) 90–93
interim prohibition 4, 11

level sensor 134–135
liquid transport 57

low permeability soils 176

manual reconciliation 32–42
mechanical pressure monitor 167–169
metal oxide semiconductor 106–107
meter accuracy 26–28, 29
metered reconciliation 37–38

National Association of Corrosion Engineers (NACE) 13
National Conference on Weights and Measures 126
National Fire Protection Association (NFPA) 23, 121, 122, 137, 150, 166
negative pressure monitor 171
nonvolumetric testing 119, 120, 149–153

observation wells
  construction 95
  design 94–95
  installation 96–97
  sampling 97–103
Occupational Safety and Health Administration 23

Petroleum Equipment Institute (PEI) 12
photo-ionization device 107
photocell 133
photometry 132, 133
piping design 11
piping tightness testing 171–173
positive pressure leak monitor 166–170
precision test 150
pressure monitor 166–171
pressure transducer 130
product level measurement 129

radiation 92
radioactive tracers 81
reference tubes 122, 128, 139, 140
remote pumpng 13–14, 165
residual materials 57
Resource Conservation and Recovery Act (RCRA) 4, 14, 23
rhodamine B 78

sacrificial anode 5, 7–8, 12
   high potential magnesium 6
   standard magnesium 6
   zinc 6
secondary containment 109–111, 174, 176
seismic refraction 81
self compensating 122, 128
soil adsorption 57
soil cements 181
soil cores 83–84
soil sampling 73
sorption/retardation 60
stabilization time 127, 138, 139, 145, 157
static reconciliation 33, 35–37
statistical reconciliation 41, 42–46
Steel Structure Paint Council 9
suction pumping 13–14, 15, 165

tank distortions 144–146, 158
tank geometry 31
temperature 25–26, 39–40, 124, 126–129, 137–140, 143, 157
temperature-compensated reconciliation 39–40
temperature-compensated testing 126–129, 138, 139
thermal conductivity sensor 99

thermistor 126, 127, 140
throughput 25, 27, 29, 41
tracer leak detection 151–153
trend analysis 122, 125, 139
trigger value 38

u-tube 107–109
   construction/installation 109
   design 108
   installation 109
ultrasonic testing 90, 92
Underwriters Laboratories (UL) 5, 6, 9, 10, 13, 31
Underwriters Laboratories, Canada 9
Uniform Fire Code (UFC) 23

vadose zone 83
vapor control 29
vapor monitoring 173, 175
vapor pockets 140–143, 157
vapor pressure 142
vapor wells 103–107
   construction/installation 103–107
   design 104
   installation 104
variables
   inventory monitoring 25–31
   tank testing 136–149, 157
vibration 146
viscosity 138
visual inspection 90
volumetric testing 119, 120, 122–126

water gauging 32
water table 52, 59, 61, 62, 143–144
well types 69–71